"A fascinating, comprehensive, and ac ⸻
most fundamental questions. If you're wondering who we are and why we're here, this book is a great place to start."

—Mark Gober, award-winning author of *An End to Upside Down Thinking* and board member at the Institute of Noetic Sciences

"A book about heaven and God that makes a strong case for both—coming from a pathbreaking astrophysicist. A mind-expanding book. Not to be missed."

—Ervin Laszlo, author of *Science and the Akashic Field*

"A gifted scientist's wonderful look at human consciousness and the hereafter through the mind-bending revelations of modern science. A beautiful achievement!!"

—Larry Dossey, MD, author of *Space, Time, and Medicine*

"Anyone who reads this book will be challenged to reframe their view of mind and consciousness in ways that offer an alternative to materialism. The table of contents alone is a brilliant scope note, and the coauthors do a beautiful job with each probing chapter. Great book, engaging writing, readable by all."

—Stephen G. Post, PhD, president of the Institute for Research on Unlimited Love

"This is a delightful and thought-provoking book, whose main author is an eminent scientist and a Christian who is highly original in his thinking. It offers some unconventional and many insightful responses to fundamental questions about the nature of reality and the amazing universe in which we live, the nature of God, and about how our deaths are unlikely to be the end but may well lead to a heaven where we are welcome."

—Eric Priest, PhD, professor emeritus at the University of St. Andrews, Scotland, and a fellow of the Royal Astronomical Society

"Bernard Haisch and Marsha Sims get it. Consciousness is causal and fundamental. And in their book *The Miracle of Our Universe* they lay out the new paradigm and what it means in a unique way. One that accurately makes the link between science and religion."

—Stephan Schwartz, author of seven books and more than one hundred technical papers, and distinguished faculty member, Saybrook University

"If you're a fan, like me, of quantum physics, Stephen Hawking, Deepak Chopra, or Eckhart Tolle, then you will love Dr. Bernard Haisch and Marsha Sims's latest book, *The Miracle of Our Universe*."

—Frances Kermeen, author of *The Myrtles Plantation*

"Science as practiced in Western universities is regrettably limited in its scope, avoiding many topics that are of interest to the intelligent layman—among them, the nature of life and death. Bernard Haisch, as a successful mainstream scientist and past editor of the wide-ranging *Journal of Scientific Exploration,* and Marsha Sims as the executive editor of the journal, are well qualified to review and comment on the relationship between scientific thinking and such ephemeral topics."

—Peter Sturrock, emeritus professor of applied physics at Stanford University

The
Miracle
of Our
Universe

The
Miracle
of Our
Universe

A NEW VIEW OF CONSCIOUSNESS, GOD, SCIENCE & REALITY

Bernard Haisch, PhD and Marsha Sims, MM

NEW
PAGE

This edition first published in 2023 by New Page Books, an imprint of
Red Wheel/Weiser, LLC
With offices at:
65 Parker Street, Suite 7
Newburyport, MA 01950
www.redwheelweiser.com

ISBN: 978-1-63748-014-4
Library of Congress Cataloging-in-Publication Data available upon request.

Cover art by Bruce Rolff/Stocktrek Images/Getty
Cover and interior design by Sky Peck Design
Typeset in Arno Pro

Printed in the United States of America
IBI

10 9 8 7 6 5 4 3 2 1

Contents

Preface

This book is written as a joint effort by astrophysicist Bernard Haisch and his long-time wife, Marsha Sims.

After he retired from a long career at the forefront of scientific research, Bernard was diagnosed with Parkinson's disease, which made it difficult for him to write or type. So Marsha stepped in to act as his co-author and scribe, and managed to turn a manuscript into a book. She also introduced important new ideas. Throughout, however, the two have worked as a team to express the beliefs they have developed and shared over a lifetime.

In this book, Bernard and Marsha have chosen to refer to God as "he," but of course the consciousness they describe is formless, genderless, non-binary, and asexual. It's only the idiosyncrasies of the English language that required them to choose a pronoun when referring to him.

A Cosmic Simulation

*I regard consciousness as fundamental. I regard matter
as derivative from consciousness. We cannot get behind
consciousness. Everything that we talk about, everything that
we regard as existing, postulates consciousness.*

MAX PLANCK

HAVE YOU EVER WONDERED WHY and how the world around you
came to exist? Have you ever asked yourself whether there may be
an afterlife awaiting you after your physical body dies? Could there
actually be a God and a heaven of some sort?

The hypothesis we present in this book is that our seemingly
physical universe of matter and energy is a *cosmic simulation*—a
virtual reality—that is brought into existence by the thoughts of
a universal consciousness that is vast beyond our wildest imagina-
tions, perhaps even infinite and eternal. We call this consciousness
God or Creator.

So why would a universal consciousness create a virtual universe? Could it be to evolve itself through the free will exercised by
the beings that inhabit that universe? Perhaps to create experiences
for itself? In this virtual universe, literally everything is consciousness in action. We and other creatures are the offspring of this universal consciousness, and our role is to be its eyes and ears.

In this universe, nothing is ever what it seems to be—just like the virtual simulations in a video game, where each player chooses an avatar from a wide range of characteristics and abilities to use during the course of the game. These avatars interact with each other and the environment, while a game master acts as the organizer, officiant, arbitrator, and moderator. In our view of the universe as a cosmic simulation, God acts as game master; we are the avatars. As we act out our lives' dramas, we are activated, not by electronics, but rather by the consciousness of God, who experiences whatever we may be experiencing. In a sense, our experiences are thus his own.

"In the beginning was the logos," we read in John 1:1, "and the logos was with God, and the logos was God." *Logos* is a Greek word that means the expression of a thought. There are many subtle different usages of the word, but its interpretation as an "expression of a thought" is particularly relevant for the God model.

We propose a model of God that starts with the premise that *at first there was nothing.* Then all things began when the logos (the Word) appeared. The logos is God and the logos is love. God, who is an infinite eternal consciousness and spirit, created the universe with the energy of love, which is not just a warm, fuzzy feeling, but actually a powerful force. Then God created a *virtual universe* of what seems like physical matter, energy, and time, booting up that simulation with a phenomenon called the Big Bang, which gave birth to light composed of photons and other particles, which in turn generated a vast field of electromagnetic energy (zero point energy, or ZPE), which is ubiquitous and infinite throughout the universe.

Galaxies were formed out of immense clouds of gas that collapsed and rotated. As they evolved, stars formed within them—massive glowing bodies of gas that emit vast quantities of light and whose temperatures range from thousands of degrees at their "surface" to millions of degrees at their core. Our star is the Sun, inside of which roughly a million earths could fit.

After billions of years, most stars merged into galaxies, like our Milky Way, which contains approximately 100 billion stars. The Hubble space telescope and the more recent NASA James Webb telescope have done deep-field imaging of galaxies from which it can be inferred that there are an unimaginable 200 billion trillion stars in the observable universe.

Astronomer Edwin Hubble discovered in the 1920s that the entire universe is expanding, which leads to a mind-boggling conclusion. If we plot a representative sample of (expanding) galaxies—say a few thousand, not a billion—the data show that space is expanding. But if we plot this data backward, it leads to a scenario in which these innumerable star-filled galaxies are actually hyper-compressed into a point in space having infinite density. And as we reach farther and farther back in time—as we approach time T=0—everything comes to a grinding halt. Here, reality becomes merely a virtual tableau in a timeless realm. And it is from this nothingness—the *Ein Sof*—that God came forth.

The Simulation Hypothesis

The idea that what we call reality is actually a super-sophisticated video game, a concept made popular by Rizwan Virk in his book *The Simulation Hypothesis,* raises the fundamental question of whether we are all actually characters existing inside a simulated reality that is so well rendered that we cannot distinguish it from "physical reality."

Consider your favorite armchair. It feels solid and stable while you're sitting in it. Yet it is anything but solid. After all, it is composed of atoms. You may recall from your science classes that an atom is often depicted as a miniature solar system with electrons orbiting a nucleus. Most of it is empty space. But according to quantum mechanics, the picture is somewhat more abstract. The electron has a form that can only be described as a "cloud of probability," not an orbiting particle. According to science writer Anne Marie Helmenstine:

> An electron cloud is the region of negative charge surrounding an atomic nucleus that is associated with an atomic orbital. It is defined mathematically, describing a region with a high probability of containing electrons.

The phrase "electron cloud" first came into use around 1925, when Erwin Schrödinger and Werner Heisenberg sought a way to describe the uncertainty of the position of electrons in an atom.

So, when you are sitting in your favorite armchair, you are actually being supported by clouds of probability. Seems a bit precarious, but this is how it is in the world of atoms. And the same applies to all other apparently solid physical matter.

But probabilities are numbers. So when you sit in your armchair, you are actually demonstrating how matter and number interact with each other. In this book, we develop the idea that a matter-based universe can be replaced by a number-based universe. And this is what leads us to the likelihood that reality is a simulation.

A simulation interpretation of reality is backed up by the fact that atomic particles—like electrons—are seen as not even existing until they are measured. Experiments in quantum physics are strengthening the evidence that what we perceive as physical reality only becomes "real" when observed. As British cosmologist and astrophysicist Martin Rees pointed out:

In the beginning there were only probabilities. The universe could only come into existence if someone observed it. It does not matter that the observers turned up several billion years later. The universe exists because we are aware of it.

This is called the *measurement problem* in quantum physics—a problem that leads to conclusions that, to most, are bizarre to the nth degree.

How is it possible that, if you cut open your armchair and take some kind of measurement of a few electrons inside—never mind what kind of measurement—those electrons will be "promoted" into existence by virtue of their being measured, while gazillions of other electrons are left to languish in a cloud of probability? If we try to explain this behavior with real, solid, permanent particles, we can't.

But if the world is a simulation, anything goes.

The Power of Modeling

In fact, if we accept the possibility of matter and numbers interacting with each other, we can develop a mathematical model that becomes an ever-better simulation of reality. This is an especially useful approach when studying problems in cosmology, as shown by Stephen Hawking and Leonard Mlodinow in their work *The Grand Design,* in which a mathematical model becomes ever more realistic and valuable.

Here is an example. Picture a test aircraft in a wind tunnel. Note all of the various responses it has as conditions are varied and test results are charted. This process can be a very useful way to study how a real aircraft will behave in actual flight. In the same way, an ever-more complicated mathematical model of the universe can also be developed so that it becomes an ever-better simulation of reality. Indeed, as physicist and astronomer James Jeans observed, as mathematical models of the universe become more refined, it begins to appear that the universe was "designed by a pure mathematician."

This book takes a similar approach. Here, we present a *model* of God—of the creation of the universe and of our purpose in life. We call this the "God model" and believe that it represents a model of God, the universe, and reality that is compatible with 21st-century science—a model that is worth *believing* in.

Throughout this book, we will address important questions about life and reality, including

◊ What kind of being could God be?

◊ What is our relationship to God?

◊ Could there be more than one God?

◊ Where could God have come from?

◊ How does the Big Bang theory figure into this?

◊ Is there an afterlife and what could it be like?

◊ What is the purpose of life?

◊ Is consciousness all there is?

◊ Is the universe a virtual reality?

◊ Why is hell impossible?

◊ How can we reconcile a belief in an all-powerful personal God with the discoveries being made in modern science?

These are pretty audacious questions, to be sure. That's why we approach them in terms of developing a model, one that is self-consistent and convincing.

Introduction to the God Model

There are a number of non-religious belief systems that question both the existence and the nature of God—from atheists who don't believe in a divine being, to agnostics who deny we can ever know anything about the creation of the universe, to deists who believe that, while God created the universe, its actions are determined by inalterable laws. But the God that we present in this book is a model that even a die-hard atheist can find worthy of serious consideration. The God of our model does not hurl lightning bolts or care one way or the other whether or not we believe in him (or her) during our life on earth. The God of our model is endlessly tolerant and infinitely loving. We can't really offend him no matter how hard we try. He loves us.

This view of God as consciousness has important implications: *God requires us in order to know himself.* Thus we have a fairly important role to play in the overall scheme of things. But who are we to make such audacious claims? And why should you believe a word of what we have to say?

The answer is that you should not *believe* anything we say. Rather, we ask that you suspend your judgment on these all-important questions until we have laid out our case—the discoveries of science, the ancient wisdom of the religions of the world, and the mystical insights of thinkers who have come before us. Then ask a different question: Does this all hang together? Does it make more sense than any other alternative? Is it more rational than what you have been taught to believe?

We maintain that logic, science, religion, and ancient wisdom—what Aldous Huxley called the "perennial philosophy"—can be combined today to yield some powerful arguments about what God and heaven and hell *cannot* be. And although knowing what something *cannot* be is not as good as knowing what something actually *is*, it's still a good place to start.

We hope you'll find our model convincing—or at least give it serious consideration. And keep in mind what biblical scholar Bart Ehrman warned:

> The search for truth takes you where the evidence leads you, even if, at first, you don't want to go there.

CHAPTER I

Near-Death Experiences

Everyone who is seriously involved in the pursuit of science
becomes convinced that a spirit is manifest in the laws
of the Universe.

ALBERT EINSTEIN

MYSTICAL EXPERIENCES HAVE BEEN REPORTED throughout human history, but it is only in recent years that medical tools and practices have been put to use to gather data about them. One of the most common of these is near-death experiences.

Reports on these experiences—called NDEs—can be found going back at least to the Roman Empire. Throughout the ages, mystics and visionaries have reported similar experiences, and there is now ample evidence to claim that NDEs have earned the right to be recognized as worthy of study alongside more traditional fields like botany or anthropology. Indeed, today there is a growing awareness among doctors and scientists of the reality of near-death phenomena and a growing willingness to research and discuss them.

Many great discoveries have been glibly asserted to have "changed everything." But in the case of research into near-death experiences, this is undoubtedly true. In his pioneering 1975 book *Life After Life*, philosopher and psychiatrist Raymond Moody investigated and defined near-death experiences and thereby opened up a whole new dialogue about the possibility of an afterlife. Likewise psychologist Kenneth Ring has identified and explored a consistent set of value and belief changes associated with people who have had near-death experiences.[1] Among these changes, he found a greater appreciation for life, higher self-esteem, greater compassion for others, less concern for acquiring material wealth, a heightened sense of purpose and self-understanding, a desire to learn, elevated spirituality, greater ecological sensitivity and planetary concern, and a feeling of being more intuitive.

There are now hundreds of books and many thousands of interviews illuminating near-death experiences. They appear to reveal visions of a spiritual world open to the experiencer, but invisible to others in attendance. And there is a marked similarity between these many reports.

Those who have experienced NDEs report sensations of floating out of their unconscious bodies and seeing themselves— perhaps on an operating table, perhaps at an accident scene, perhaps in bed at home or at a hospice—from a position above where their bodies lie. They describe amazing "life reviews" in which all the tiny details of their physical lives flash by and are examined. Time races by and yet seems to stand still. But most important, they report an intense feeling of homecoming—of returning at last to a long-forgotten place. Their earthly lives start to fade away as if they were waking from a dream. They report encountering their true spiritual selves, which consist of many previous lives.

Most noteworthy of these sensations is that of returning to a true spiritual home and feeling that "everything is more real than real."

We take these reports seriously.

---

1   Kenneth Ring, PhD in social psychology, University of Connecticut, co-founder of the International Association for Near-Death Studies (IANDS).

Pioneers in NDE Research

American psychiatrist George Ritchie had a near-death experience at age twenty when he was pronounced dead twice by a doctor. In the nine minutes during which he was claimed to be dead, he experienced many realms of heaven, passed through several dimensions of time and space, and had a vision of Jesus Christ. He reported his experiences in his best-selling book, *Return from Tomorrow* (1978). In fact, it was Ritchie's experience that inspired Raymond Moody to write his landmark book *Life After Life*, in which he coined the term "near-death experience." This book opened the door to scientific investigation of spiritual realms. Moody went on to document over 150 case studies of people who were revived or spontaneously came back to life after being pronounced medically dead for several minutes.

Moody's work prompted a stream of books, articles, films, and organizations that focused on near-death phenomena. From 1975 to 2000, almost 3,500 reports were investigated. There is now a veritable flood of first-hand accounts by people who have experienced another realm of existence. Reliable estimates are hard to find, but it is now believed that roughly one in ten people who come close to dying has a near-death experience.

Another important pioneer in NDE research was Elizabeth Kübler-Ross, a Swiss-American psychiatrist whose 1969 book *On Death and Dying* explored what the dying have to teach doctors, nurses, clergy, and their own families. It was here that she first discussed her theory of the five stages of grief. Kübler-Ross's work was then reinforced by the research of psychiatrist and neuro-behavioralist Bruce Greyson, who has studied near-death experiences for more than forty-five years. Greyson has become one of the world's leading experts on the science and significance of these phenomena.[2]

Dutch cardiologist Pim van Lommel became intrigued by the number of his heart-attack patients who reported near-death experiences, so he designed a research study that lasted over twenty years. In 2001, Lommel and his collaborators published their results in the prestigious medical journal *The Lancet*. Their work was hailed as the first scientifically rigorous study of NDEs.

In fact, a growing number of doctors have written about their own near-death experiences. In her book *To Heaven and Back: A Doctor's Extraordinary Account of Her Death, Heaven, Angels, and Life Again,* orthopedic surgeon Mary Neal tells of a kayaking accident in South America that took her to heaven—and back. Her book became a *New York Times* best-seller with over a million copies sold.

While on holiday in southern Chile, Neal was trapped underwater in the powerful flow of a waterfall and the currents around it for twelve minutes. Before the medical team sent to rescue her was able to resuscitate her, she saw visions of heaven and angels. By the time she was resuscitated, she had gone without oxygen for over half an hour. This near disaster transformed her life and redefined the strength of her faith.

From the combined reports of these researchers, we can begin to define the shared characteristics of near-death experiences. These include

◊ An awareness of being dead.

◊ A sense of peace, well-being, and removal from the world.

◊ Positive emotions.

2   After: A Doctor Explores What Near-Death Experiences Reveal about Life and Beyond, Irreducible Mind (co-author), The Handbook of Near-Death Experiences, (co-editor).

◊ Perceptions of the body from an outside vantage point with 360-degree vision.

◊ The ability to see and hear the activity of a medical team delivering care.

◊ Moving into and through a dark tunnel with a brilliant light at the end.

◊ Immersion in a powerful being of light who communicates telepathically.

◊ Intense feelings of unconditional love and acceptance.

◊ A minutely detailed life review that compresses years into seconds.

◊ Entering a realm of sights and sounds of incredible, unearthly beauty.

◊ Sensing an overwhelming presence of love and extreme peacefulness.

◊ Meeting long-lost friends and relatives.

◊ Perceiving everything as more real and familiar than life on earth.

◊ A feeling that we are connected to all that is.

◊ Coming to a boundary where a choice must be made about whether to stay or return to earth.

◊ Feeling a greater appreciation for life, higher self-esteem, greater compassion for others, and less concern for acquiring material wealth.

◊ A heightened sense of purpose and self-understanding, elevated spirituality, greater ecological sensitivity, and increased planetary concern.

And for some, a near-death experience is like a trip to heaven and a reminder of who they truly are.

Voices from the Edge

People often describe the profound effect that near-death experiences have on them, causing them to re-examine their lives and morals. One doctor tells of his patient's experience following an automobile accident that left him in a coma. While in this state, he had a vivid sensation of leaving his physical body and entering into darkness. He reports having the feeling of moving up through a dark tunnel toward a point of light. Suddenly, a being "filled with love and light" appeared to him and guided him through a life review. He felt bathed in love and compassion as he reviewed the moral choices he had made in his lifetime. He suddenly understood that he was an important part of the universe and that his life had a purpose.

Some who approach the boundary of death and return tell of traveling through walls to a waiting room where they see their relatives and friends. One patient reported traveling through a wall and seeing her young daughter wearing mismatched plaids, which was highly unusual for her. Another woman traveled through a wall and overheard her brother-in-law in the hospital waiting room talking to a business associate in a very derogatory manner, a conversation she was able to repeat back to him later.

One of the most dramatic experiences was reported by neurosurgeon Eben Alexander. In his best-selling book *Proof of Heaven*, Alexander recounts his own descent into a deep coma due to bacterial meningitis from a vicious strain of E. coli. Doctors estimated his chance for recovery at less than 10 percent at the outset, then less than 2 percent after a week in a comatose state. Even if he were to come out of his coma, they predicted, he would spend the rest of his life in a nursing home. The neocortex of his brain had completely shut down and he was effectively dead. He had no functioning brain.

Yet instead of dying, Alexander spontaneously emerged from his coma after eight days—and with an amazing story to tell. In describing his experience while near death, he said: "I believe we are on the greatest revelation in human thought in all of recorded history, a true synthesis of science and spirituality." His observations super-charged interest in NDE phenomena.

Prior to his own experience, Alexander had thought of himself as a scientific skeptic, one who casually dismissed accounts by his patients of NDEs. But after his own experience, he declared:

> Like many other scientific skeptics, I refused to even review the data relevant to the questions concerning these phenomena. I prejudged the data, and those providing it, because my limited perspective failed to provide the foggiest notion of how such things might actually happen . . . [But] those who assert that there is no evidence for phenomena indicative of extended consciousness, in spite of overwhelming evidence to the contrary, are willfully ignorant. They believe they know the truth without needing to look at the facts.

Alexander studied his own medical charts and came to the conclusion that he had been in such a deep coma during his own NDE, and his brain had been so completely shut down, that the only way to explain what he felt and saw was that his soul had indeed detached from his body and gone to another dimension. His experience led him to believe that angels, God, and the afterlife were all as real as could be.

By contrast, Alexander lamented, current neuroscientific hypotheses about the possible biochemical nature of memory and consciousness are all over the map, with nothing even remotely resembling a consensus in sight. He ended by formulating a new approach to consciousness:

> My new view, that is emerging in neuroscience and philosophy of mind, is that soul/spirit exists, and projects all of apparent physical reality from within itself. The brain is more a prison from which our conscious awareness is liberated at the time of bodily death, enabling a robust afterlife that also involves reincarnation. Our choices matter tremendously, and thus free will is a crucial component.

And this leads us to the foundational principle of the God model—an infinite and eternal consciousness that constructs what we perceive as reality.

CHAPTER 2

Consciousness and Reality

Advances are made by answering questions.
Discoveries are made by questioning answers.

BERNARD HAISCH

ADVANCES IN SCIENTIFIC KNOWLEDGE are now strengthening the evidence that reality takes place only when observed. This is called the "measurement problem" in quantum physics.

Human bodies are made of cells, approximately 30 trillion of them. But our bodies replace their cells throughout our lifetimes. In fact, on average, each cell is destroyed and replaced by a new cell every ten years. But from the point of view of consciousness, this raises a profound conundrum: Who are we? If all the physical matter of which we are made has been replaced, what is our relationship to the person we were ten years ago, or twenty years ago? How can we claim to be one continuous entity from birth to death if our physical bodies are constantly renewing themselves? What do we have left?

From the point of view of physics, human consciousness *somehow* emerges from the body's cells, probably our brain cells. This consciousness *somehow* makes or creates some kind of replica of an outer world and an inner world. Together, these two "worlds" create an inner experience of thought that is taken to be all that we are. This is the physicalist view—or perhaps the physicalist dogma.

But is this really a true picture of how our perceptions of reality are constructed?

We argue for a different concept of reality—one in which there is room for both a transcendent Creator and for science. This is what we call the God model.

Science and the God Model

In the God model, the theory of evolution is necessary to understand the nature and purpose of God. In this view, there exists an infinite and eternal consciousness that is *all there is*. Through this consciousness, God seeks to express himself in all his infinite creative potential. Our reality—the world around us—is one such creation. Since God is all there is, all living things on earth and elsewhere are manifestations of him. In the case of humans, we are the senses by which God is able to know himself.

Many of the world's religious traditions tell us that the world around us is an illusion created for our benefit. This is particularly true in the Eastern traditions of Buddhism and Hinduism, which explicitly tell us that the world we see is *maya*, or illusion. Even Western religions claim that this world (the "here") is different from the other, eternal world (the "hereafter"). But today, advancing science is discovering concepts that could shed light on these spiritual claims about other realms beyond the physical, thus allowing us to develop a better notion of what God and heaven and hell may be. And this model—which includes notions like near-death experiences, past lives, an afterlife, and out-of-body experiences—can lead us to a new and richer view of the role of consciousness in constructing reality.

Pioneering psychologist and philosopher Carl Jung got a glimpse of this new view of consciousness in 1944 following a major heart attack. Here's how he described his own near-death experience in his autobiography, *Memories, Dreams, Reflections*:

The beginning of 1944 I broke my foot, and this misadventure was followed by a heart attack. In a state of unconsciousness, I experienced deliriums and visions which must have begun when I hung on the edge of death and was being given oxygen and camphor injections. The images were so tremendous that I myself concluded that I was close to death. My nurse afterward told me: It was as if you were surrounded by a bright glow. That was a phenomenon she had sometimes observed in the dying, she added.

I had reached the outermost limit, and do not know whether I was in a dream or an ecstasy. At any rate, extremely strange things began to happen to me.

It seemed to me that I was high up in space. Far below I saw the globe of the Earth, bathed in a gloriously blue light. I saw the deep blue sea and the continents. Far below my feet lay Ceylon, and in the distance ahead of me the subcontinent of India. My field of vision did not include the whole Earth, but its global shape was plainly distinguishable and its outlines shone with a silvery gleam through that wonderful blue light. In many places the globe seemed colored, or spotted dark green like oxidized silver. Far away to the left lay a broad expanse—the reddish-yellow desert of Arabia; it was as though the silver of the Earth had there assumed a reddish-gold hue.[3]

---

3 Jung's NDE and visuals of Earth from space occurred in 1944. It is interesting to note that we have a photo taken from an Apollo mission about 1970 (taken about 36 years later) hanging on our wall that is very similar in appearance.

Then came the Red Sea, and far, far back—as if in the upper left of a map—I could just make out a bit of the Mediterranean. My gaze was directed chiefly toward that. Everything else appeared indistinct. I could also see the snow-covered Himalayas, but in that direction, it was foggy or cloudy. I did not look to the right at all. I knew that I was on the point of departing from the Earth. Later I discovered how high in space one would have to be to have so extensive a view—approximately a thousand miles! The sight of the Earth from this height was the most glorious thing I had ever seen. This experience gave me a feeling of extreme poverty, but at the same time of great fullness. There was no longer anything I wanted or desired. I existed in an objective form; I was what I had been and lived. At first the sense of annihilation predominated, of having been stripped or pillaged; but suddenly that became of no consequence.

This experience led Jung to admit that we can never completely understand death, the greatest human mystery.

Since death lies so far beyond our intellectual comprehension, Jung concluded, we must engage it with our souls. Death demands a leap of faith. It challenges us to lay claim to something unknown and unknowable. It asks us to dive into the great adventure ahead of us. He concluded:

Not to have done so is a vital loss. For the question . . . is the age-old heritage of humanity: an archetype, rich in secret life, which seeks to add itself to our own individual life in order to make it whole.

And this "secret life" of consciousness—that infinite and eternal consciousness that is the foundation of the God model—is what constructs the world we see around us and the role we play in it.

So what does that mean for us as we contemplate what happens to us after we die? If, in fact, we persist as a part of this infinite consciousness, is there another life in store for us after death? Many traditions claim there is. And that's where we'll turn our questions next.

CHAPTER 3

Reincarnation and Past Lives

*It's not that I'm afraid to die; I just don't want to
be there when it happens.*

WOODY ALLEN

Yet, despite Allen's quip, it turns out that you may actually want
to be there . . . at least in spirit.

The concept of reincarnation is grounded in a belief in the
rebirth of souls. According to spiritual traditions that believe
in reincarnation, souls living in physical bodies that die go to an
afterlife, where they are trained and reprogrammed by higher spiritual beings. After some time, they are reborn into a new body and
return to earth with a new life plan. This process involves a kind of
spiritual evolution that is, to some extent, analogous to Darwinian evolution, spanning multiple lifetimes and multiple realms of
existence.

Research into the possibility of reincarnation has found some
convincing evidence for it. Psychiatrist Ian Stevenson became
known for his research into cases he considered suggestive of
reincarnation. After studying 3,000 cases of children who claimed
to remember past lives, he concluded that emotion, memories,
and even physical bodily features could be transferred from one
life to another. Stevenson helped found the Society for Scientific

Exploration in 1982 and authored over 300 papers and books on reincarnation.

Commenting on Stevenson's research, renowned astronomer and astrophysicist Carl Sagan said:

> There are three claims in the [parapsychology] field which, in my opinion, deserve serious study with the third being that young children sometimes report details of a previous life, which upon checking turn out to be accurate and which they could not have known about in any other way than reincarnation.

Past Lives

As managers of the scholarly publication *The Journal of Scientific Exploration*, we got to know Stevenson quite well as a very credible detail-oriented scholar. His articles included reports of children with malformed or missing fingers who said they recalled the lives of people who had lost fingers. In another example, one boy, who had birthmarks that resembled entrance and exit wounds of a bullet, said he recalled the life of someone who had been shot. Another child with a scar around her skull three centimeters wide said she recalled the life of a man who had had skull surgery.

Stevenson conducted detailed interviews with these children—all under five years of age—and interviewed their parents, friends, and relatives. After researching the memories the children had of their former lives, he went to the locations where these deceased persons had lived and interviewed their relatives and friends. He even went into police records to get information. In many of the cases, he found that the witness testimony and autopsy reports appeared to support the existence of injuries that matched the scars and birthmarks on the children he studied.

In the case of the young boy who recalled the life of someone who had been shot, the sister of the deceased told Stevenson that her brother had shot himself in the throat. The young boy showed Stevenson a birthmark on his throat. Suspecting that he might also have a birthmark on the top of his head where the bullet had exited, he looked and found one underneath the boy's hair.

When Stevenson retired in 2002, his work was continued by pediatrician and psychologist James Tucker. Tucker has carried on Stevenson's studies of reincarnation in the most methodical and rigorous manner, and the data from his research presents another compelling, objectively measured case for the continuation of consciousness.

In the intriguing case of James Leininger—an American child who, at age two, began having intense nightmares of a plane crash—Tucker uncovered facts that further confirmed his belief in past lives. From age two, Leininger had played with nothing but planes. When he was three, his parents watched as he went over a plane as if performing a preflight check. When his mother bought him a toy plane and pointed out what appeared to be a bomb on its underside, Leininger corrected her and told her it was a drop tank. He once told his father that he flew a Corsair, and that "they used to get flat tires all the time." The boy told his parents that his plane had been shot down "by the Japanese."

Tucker reported that the child's memories included being an American pilot named James Huston, who was killed when his plane was shot down by the Japanese in World War II.[4] These memories included details of an American aircraft carrier, the first and last names of a friend who was on the ship with him, the location of the attack that killed him, the nature of the hit that

4   Jim B. Tucker, M.D., *The Case of James Leininger: An American Case of the Reincarnation Type,* Division of Perceptional Studies, University of Virginia.

brought down his plane, and other specific information involved in the fatal cash.

All these details were later confirmed to be accurate. Leininger's parents eventually discovered a direct correspondence between their son's statements and the death of an American pilot named James Huston who was shot down in World War II in the place their son described and in the way he described. Many people—including those who knew the fighter pilot—think the boy is James Huston reincarnated.

All this leads us to ask a key question: If what we have learned about near-death experiences and what we have learned about consciousness and what we have learned about reincarnation is all true—what does that mean for the existence of an afterlife? And by afterlife, do we really mean *after we die*?

Yes, we do. And although some of these questions about human existence may seem perplexing, all we ask is that you keep an open mind.

CHAPTER 4

The Frontier of Science

Consciousness cannot be accounted for in physical terms.
For consciousness is absolutely fundamental. It cannot be
accounted for in terms of anything else.

ERWIN SCHRÖDINGER

WHEN BERNARD WAS A FULL-TIME astrophysicist, he didn't have much time for spiritual matters. This same situation applies to most career scientists. They don't have time for anything else when sixty-hour work weeks are barely enough to accommodate getting that half-finished paper written, or keeping up with the latest journal articles, or preparing for a fast-approaching conference deadline or the deadline for the next round of highly competitive grant proposals that keep the research ball rolling. When that newest request for proposals is issued, anxiety begins to rise. No doubt about it; the life of a professional scientist is a busy one.

In more recent years, however, Bernard has devoted his life to exploring the nexus between science and spirituality. As a scientist, he carried out research in astrophysics, but at this stage of his life, his fascination with matters spiritual has overtaken his scientific curiosity and he has turned his thoughts to a belief in God. Does he believe in God? The quick answer is *yes*. He has great confidence in the existence of a Creator. (Marsha does as well.) But not some impersonal God who's lurking somewhere in the background and does not really matter much. On the contrary, both believe in a very *personal* God, one to whom it is even worth praying.

And along with a belief in God come some intriguing questions. Is there an afterlife? Is there a heaven? Is there a hell? As we shall see, the classical model of heaven has some real problems. Is this a place where you would actually want to stay for any length of time? And as for the classical model of hell, the application of a little elementary logic shows that the existence of a classical hell is . . . well, impossible. Here, we present a model that overcomes these problems and offers an audacious new view of the afterlife.

In his book *The God Theory*, Bernard argued that science and spirituality, when properly viewed, do not have to be at odds with each other. In fact, they can be mutually reinforcing. Many of his professional colleagues saw this as a heretical claim. But the more time he spent reading and writing and just plain pondering fundamental questions—like how the universe came to be—the more convinced he became that he was on the right track. And surprisingly—or perhaps not so—his book struck a resonant chord with a public starved for answers to some of life's most perplexing questions.

In *The God Theory*, Bernard wrote:

> Let me make it clear that I don't claim to speak directly to God. I am too conditioned as a scientist for that. In fact, if God ever calls, my line will probably be busy . . . but he might try my email.

Not surprisingly, Bernard is still not receiving regularly scheduled messages from God, but he has learned that, when God speaks, he does so in a still, small voice, as Sir Arthur Eddington noted.

Since turning his attention more fully to spiritual questions, Bernard has become more open to divine guidance prompting him to hear and see things that he would never have bothered to notice before. As a result, he has become more comfortable with the conclusions he has drawn about God and the purpose of life as he nears the end of his own life—this one, at least. And he hopes to contribute something helpful to mankind by offering a credible model of reality that can rattle the mental cages of those who have difficulty stepping outside of the prevailing divisive ideas about God and spirituality. And remember—what we offer here is a *model* of reality, not reality itself.

Life on the Frontier

The frontiers of science move forward slowly but surely. It is said that there are three stages on the way to a new scientific idea being accepted. First, it is ridiculed. Then it is fiercely attacked and denounced. And finally, it is considered to be self-evident. That was the case for the Darwinian theory of evolution.

An Anglican bishop who had clearly not advanced beyond the first stage remarked upon hearing of Darwin's theory: "Descended from apes! My dear, let us hope that it is not true; but if it is, let us pray that it will not become generally known." His comment echoes the fear that Darwin's theory would threaten to unravel the entire social fabric.

During the second stage, Darwin's theory was subjected to attack from both scientists and religious figures in an attempt to maintain the intellectual and spiritual status quo. Today, we take evolution for granted and have difficulty understanding how anyone can deny its veracity. Indeed, Stephen Hawking wrote:

> We are just an advanced breed of monkeys on a minor planet of a very average star. But we can understand the universe. That makes us something very special.

Bernard takes Darwin's theory a step further still, making it a part of a larger narrative that involves a new take on consciousness. To him, evolution explains a great deal about who and what we are.

The world is in quite a precarious position today thanks to the weaponization of technologies that should benefit, not endanger, mankind. As theoretical physicist Richard Feynman observed: "To every man is given the key to the gates of heaven. The same key opens the gates of hell. And so it is with science." We are threatened by suitcase-size atomic bombs, by drones that can deliver all sorts of nasty things, and by religions that create martyrs and an endless parade of fanatic murderers. As Gilbert and Sullivan said: "Things are topsy-turvy."

What's missing, according to Bernard, is a view of life that combines the endeavors of science with the promise of spiritual meaning. That's not to say that he wants everyone to believe in God. In fact, he states outright that there is nothing wrong with being an atheist. He has no doubt that God does not give a hoot whether someone lives life as an atheist. He believes that each of us will enter into the same kind of afterlife when we die. And while this may seem confusing to those with no expectation of an afterlife, he is convinced that a very pleasant discovery awaits them.

Moreover, this is more than a brash opinion on Bernard's part. For years, he has kept one foot in the realm of mainstream astrophysics and the other balanced on the frontier of unorthodox science. As a result, he has been exposed to lots of heretical worldviews, some of which were way over the top. He has heard a lot of claims that the world seems to be changing (even changing rapidly) toward a more open, less hostile view of expanding consciousness. But scientists in general are remarkably resistant to a spiritual view of the world and of life. This became abundantly apparent a few years ago when he tested how open his esteemed colleagues might be to a bit of consciousness expansion.

Bernard was chairman of a major conference of the International Astronomical Union on the topic of *Solar and Stellar Flares* hosted by Stanford University. It was attended by leading experts from around the world. Bernard decided to engage cultural historian Morris Berman as speaker at an evening banquet. In his book *The Reenchantment of the World*, Berman considered how science had acquired its controlling position in the consciousness of the West. Fascinated by Berman's critique of the scientific revolution vis-a-vis a holistic worldview, Bernard anticipated providing his colleagues with some provocative fare at the banquet.

But *The Reenchantment of the World* was not light reading, and Berman's views on the role of science in the development of Western culture proved to be a bit more than provocative. In his book, he presented a perceptive study of the role of science and a cogent and forceful challenge to its supremacy. He argued that science destroyed the holistic and animistic traditions of earlier ages—traditions that had viewed man as a participant in the cosmos, not as an isolated observer. This holistic worldview must be revived, he claimed, before we destroy our society and our environment. He propounded a new worldview appropriate to the modern era, one grounded in a real and intimate connection between man and nature. That was a lot to ask the mainstream scientists at the banquet to digest.

How was Berman's talk received?

Not well.

For Bernard, on the other hand, Berman's argument played right into his own developing views on the urgent need for a worldview that acknowledged the role of an infinite and eternal consciousness from which reality and our role in it flowed. What Bernard sought was a worldview that was grounded in science, but not dominated by it. A worldview that accounted for the constructive role of spirit.

He found answers in the developing fields of quantum physics.

CHAPTER 5

The Zero Point Field

The atoms or elementary particles themselves are not real; they form a world of potentialities or possibilities rather than one of things or facts.

WERNER HEISENBERG

THE ZERO POINT FIELD (ZPF, also known as the Quantum Vacuum Field, QVF) is now broadly recognized as a vast field of electromagnetic energy. Along with its attendant zero point energy (ZPE), this field represents the underlying energy that is ubiquitous throughout the universe, even where there is otherwise nothing but a vacuum. The electromagnetic energy of this field is composed of a combination of every frequency (or wavelength) that exists—some very long, others incredibly short. This spectrum of frequencies is absolutely smooth, and mathematically increases to the fourth power. In an exactly defined mathematical sense, it is perfectly random.

To understand the history of how this field was discovered, let's back up a step—to the ground-breaking work of physicists like Max Planck, Albert Einstein, and Otto Stern.

The sun, as the brightest object in the daytime sky, emits intense "blackbody radiation," a spectrum of light generated by any sufficiently heated object. Common examples of this include the heating element of a toaster or an electric stove burner. If you turn the burner up high enough, the reddish light you see is blackbody radiation. When you sit in the sunlight on the beach, you feel the heat (the blackbody radiation) from the sun and may even get sunburned.[5]

The physics of blackbody emissions was a hot topic in 1900. On the first day of the new millennium, German theoretical physicist Max Planck made a major discovery. Contrary to the received wisdom, he found that light (blackbody radiation) was not composed of waves as assumed, but rather of streams of particles that came to be called photons. Moreover, it became clear that, in addition to blackbody radiation, a much more energetic radiation field was paired with the new component. This field came to be called the zero point field—ZPF.

In 1911, Planck integrated the theoretical concept of ZPF into his previous thoughts on energy, which were then developed by a number of other physicists, including Walther Nernst, Albert Einstein, and Otto Stern.[6] A closer examination of Einstein's own theory of photons revealed the existence of radiation that was ubiquitous and incessant. In 1913, Einstein and Stern took this a step further by commenting: "The field [ZPF] is the *only* reality." Science has demonstrated during the century since then that zero point energy (ZPE) is quite authentic.

The quantity of the energy in this field is enormous, estimated conservatively to be the equivalent of more than the total annual energy output of the sun, packed into one cubic centimeter! The

---

5   The very same kind of blackbody radiation coming from the sun can easily be reproduced in a laboratory, but of course with an enormous attenuation (turning down the brightness), so as not to set fire to the laboratory.

6   The zero point energy makes no contribution to Planck's original law, as its existence was unknown to Planck in 1900. The concept of zero point energy was developed by Planck in Germany in 1911 as a corrective term added to a zero-grounded formula developed in his original quantum theory in 1900.

prospect of diverting a tiny fraction of this energy for human purposes would represent a breakthrough on a par with the discovery and worldwide use of electricity.

Heretofore, the possibility of any such breakthrough has been dismissed because it is seen as trying to capture electromagnetic energy from below the quantum vacuum ground state, which we agree is impossible. However, there is a force called the Casimir force that may provide the answer. This force occurs when two metal plates are pushed together by an over-pressure of the ZPE from the outside. This creates a space known as the Casimir cavity that can take advantage of this force by allowing sub-ground-state energy levels.

If this seems more than you can comprehend, don't worry. As theoretical physicist Wolfgang Pauli once quipped: "The best that most of us can hope to achieve in physics is simply to misunderstand at a deeper level."

Modern broadcasting technologies (radio, television, the Internet, etc.) have enhanced communication around our planet. Our world is full of their output of electromagnetic radiation, which has nothing to do with radioactivity, just as zero point energy has nothing to do with radioactivity. But apart from radioactivity, we know that all matter is stable. And we know that the electrons in atoms provide the structure of matter.[7] The question is, what is there to keep the electrons in their orbital configurations?

In the view of classical physics, point-like electrons literally orbit the nucleus of an atom. In the view of modern *quantum*

---

7   It was proposed by Rutherford in 1911 that the electron is a point-like object with a negative charge orbiting a dense nucleus with a positive charge, much as the planets orbit the sun. Unfortunately, the laws of electrodynamics dictate that an orbiting charged object must lose energy. This loss of energy would force the electron to spiral into the nucleus and be annihilated in less than a billionth of a second. In 1923, Bohr refined the Rutherford model by specifying that only certain jumps between electron energy levels can be allowed. In retrospect, this was more fiat than physics.

physics, however, they reside in spread-out stationary orbitals. But if we combine the classical-physics concept of a hydrogen atom with the concept of zero point energy, we find a remarkable fact. For any given electron, there is a balance between how much energy is emitted and how much energy is absorbed. And this strongly suggests that electrons are stabilizing atoms throughout the cosmos. In this view, it is the ZPF/electron interactions that are providing the stability of the entire universe. This suggests that the ZPE is involved in the creation process of our virtual reality.

The energy output of a thimbleful of zero point energy is almost impossible to comprehend—enough energy to power the Big Bang. In fact, we don't even know where in the frequency spectrum the zero point field ends—or if it is infinite. And yet ZPE is an intrinsic and unavoidable part of quantum physics. It has been studied, both theoretically and experimentally, since the discovery of quantum mechanics in the 1920s, and there can be no doubt that zero point energy is a real physical effect.

In the discipline of physics, the terms zero point field and quantum vacuum refer to the same vast thing—a vast field of energy that fills the universe everywhere. But that energy field is neither zero nor a vacuum. It is, in fact, the greatest reservoir of energy in the cosmos. And although we can't see it, its existence is well established.

A vast, invisible, universal pool of energy? This looks and smells like another one of those mind-bending quantum paradoxes of which physicists are so fond. And to explain what generates this vast pool of energy, we have to use quantum logic, not ordinary everyday reasoning. For this, we'll turn to something called the Heisenberg uncertainty principle.

The Heisenberg Uncertainty Principle

The Heisenberg uncertainty principle states that we can never know both the position and speed of a particle—like a proton or an electron—with perfect accuracy.[8] In fact, the more we know about a particle's position, the less we know about its speed, and vice versa.

Imagine a race car crossing the finish line at a racetrack. Now ask yourself exactly how fast the car was going at the precise instant it crossed the finish line. In the macroscopic world—that is, in our physical world of objects—we can measure the speed with what appears to be arbitrary precision. But in the quantum world— where we ourselves, and the car, and the track, and the measuring device would be smaller than an atom—things are very different.

In the quantum realm, the harder we push our equipment to yield a precise velocity reading, the less precise the position on the finish line will be. This is not due to any imperfection in the equipment or to an inept technician. This behavior is actually a law of nature that there's no getting around.

So what does this have to do with the origin of the zero point field and the quantum vacuum?

Imagine a very narrow laser beam pointing in a given direction. Next, picture this laser beam as consisting of photons that all have the same frequency (or wavelength). Then slowly shut the power down on your laser generator. Common sense tells you that, when the laser generator is off, there will be no laser beam at that direction or frequency. But here's where the Heisenberg principle comes into play.

---

8 Werner Heisenberg was a German theoretical physicist and one of the key pioneers of quantum mechanics. He published his work in 1925 in a breakthrough paper when he was only twenty-three years old.

The uncertainty principle says that nothing can be *absolutely* reduced to zero, because particles at that level pop into and out of existence. So the minimum number of photons streaming in that direction at that frequency will be one-half of a photon. But this immediately raises another red flag. Who ever said that photons come in half-sizes like shoes?

Let's try an alternate interpretation. Maybe there is a whole photon present, but it is only present half the time. Photons that are only present for a short interval of time and then disappear are okay. They don't violate the principle. By extension, anything that is forced to exist also falls within the principle, but it will manifest with half its natural "strength."

But is this really a better explanation?

To answer that question, let's take this one step further and look at something called the Higgs field—a phenomenon that cannot be directly measured, but whose effects can. And to do this, we have to consider two of the toughest problems in physics—the problem of inertia and the problem of mass.

CHAPTER 6

The Problem of Inertia

Not only is the universe stranger than we imagine,
it is stranger than we can imagine.

SIR ARTHUR EDDINGTON

TWO OF THE DEEPEST MYSTERIES in physics are the origins of gravity and the source of inertia. Gravity is the force that attracts a body toward the center of the earth, or toward any other physical body having mass.[9] Objects having mass have a property called inertia, which is the tendency of matter to continue in its existing state of rest or uniform motion in a straight line unless that state is changed by an external force. You can also regard inertia as a resistance to acceleration. (In fact, that's the correct way to picture it.)

But inertia is not a property that is intrinsic to matter or to the particles that make up matter. Particles acquire this resistance to acceleration from one of two sources: the effects of a Higgs field, discovered by British theoretical physicist Peter Higgs, or the zero point field.

---

9   Aristotle believed that objects in motion need assistance to stay in motion. He was wrong.

The Higgs Field

Inertia is such a critical property of matter that the science of physics has long been stymied by its lack of understanding of it. After all, we are talking about *mass* here and, if you want to have mass, you need inertia. Then in 1964, three teams proposed related but different approaches to explain how mass arose. They proposed a new field of energy that exists in every corner of the universe. This came to be called—somewhat unfairly given the other authors—the Higgs field. This energy field is accompanied by a fundamental particle that came to be called the Higgs boson. And while the Higgs field cannot be directly measured, the accompanying particle can, and this, the researchers claimed, is the source of inertia.

Finding the Higgs boson thus became the primary goal of the Large Hadron Collider (LHC), a massive underground particle accelerator seventeen miles in circumference that straddles the French-Swiss border near Geneva. It was built by the European Organization for Nuclear Research (CERN) between 1998 and 2008 in collaboration with over 10,000 scientists, hundreds of universities and laboratories, and more than 100 countries at an estimated cost of about 5 billion US dollars.

In this vast accelerator, two beams of protons are accelerated in opposite directions until they are moving at nearly the speed of light—fast enough to circle the track more than 11,000 times a second. When the two beams collide head-on, all hell breaks loose. On impact, many new particles are created, some lasting less than a billionth billionth millionth of a second (the top quark). Sorting through the millions of particle tracks using artificial intelligence algorithms, scientists are able to recognize the "signatures" of the particles created.

In 2012, after a forty-year search, the Higgs boson was finally considered to be identified when the number of detections hit the agreed-upon statistical level of one chance in a million that the detections were statistical flukes. The new particle, named the Higgs boson, was subsequently confirmed to interact with particles that are massless. But why is this important to us here?

Higgs Field or Zero Point Field?

In modern particle physics, it is assumed that all of space is uniformly filled with an invisible substance that's sort of like molasses. When a particle, like an electron, tries to move through this molasses, the resistance it encounters is what we interpret as the mass of the particle. In fact, the idea is that different particles have different degrees of stickiness, which means they experience a different amount of resistance as they try to burrow through this pervasive substance. But the discovery of the Higgs boson promises to rewrite the very meaning of nothingness, because the field, or atomic molasses, is essentially a ubiquitous and consistent occupant of space.

So, if the Higgs field and its accompanying particle are now confirmed, is there room for the zero point field?

The answer is yes, for several reasons—some more complicated than others:

◊ Not all of the mass of protons can be attributed to the Higgs field. The kinetic energy of the gluon and the quark account for nearly 70 percent.

◊ Physics has no need for new and abstruse particle properties, since the ZPF theory is based on *good ole electromagnetism.*

◊ The Higgs mass is 10^{17} times smaller than the Planck mass (a fundamental constant), but quantum corrections from Higgs interactions with other particles should cause the two masses to be nearly equal. So, there seems to be a contradiction here that has come to be known as the "electroweak hierarchy" problem.

Now, remember Wolfgang Pauli's warning and don't worry if you don't understand all this. In fact, all of this needs further investigation. There are still important issues to deal with here—not the least of which is why we feel the effects of inertia on our bodies when we are accelerating or decelerating, but not when we are at rest or in constant motion.

Imagine you are in a jumbo jet taking off from the runway. As the plane accelerates, you are pushed backward in your seat as you experience either the zero point field or the resistance of the Higgs field. Once you are in flight, however, those effects disappear and you can easily move around in the cabin. But this all changes as you prepare to land. As the plane decelerates, you are pushed forward in your seat either because of the ZPF, or because the Higgs field is flowing past you in the opposite direction of the landing.

But didn't Einstein explain all this with his general theory of relativity, which purported to explain mass and gravity? And didn't Eddington prove that general relativity is correct with his observations of the 1919 eclipse? True enough. But then in 1989, Harold Puthoff published the article "Gravity as a zero-point fluctuation force" based on work by Soviet dissident and Nobel Laureate Andrei Sakharov, which models gravity, not as a fundamental force, but rather as a secondary effect of the zero point field. Wow!

This introduced a whole new angle on what gravity could be and where it could come from. Is it possible that the zero point field—that vast field of energy that fills the universe—is also the origin of gravity, one of the fundamental forces of physics? If so, this suggests a deep new insight into fundamental physics—that gravity might prove to be an electromagnetic process.

The Problem of Mass

But that just raises another question. What is the origin of *mass?*[10] Mass is what everything appears to have, including you and me and everything else on the planet. And this implies that mass is just a property that comes along with everything—from a dust particle floating lazily through the air to a 500-ton, fully loaded jumbo jet.

But one form of mass is inertial mass. Toss a twenty-pound medicine ball to (not *at*) your fitness partner, then toss a one-tenth-ounce ping-pong ball and it becomes immediately evident that the medicine ball has a lot more inertial mass than the ping-pong ball—3,200 times as much, to be exact.

But inertial mass is not an intrinsic property of matter. It appears when matter is accelerated or decelerated. Galileo, Newton, and Descartes all wrestled with the problem of what causes inertial mass.

According to the standard model, protons and neutrons are made of quarks. So, we might be tempted to conclude that the mass of a proton or neutron resides in the masses of the quarks of which they are composed. But we'd be wrong again. Although it's quite difficult to determine the precise mass of quarks, they are substantially smaller and lighter than the protons and neutrons they

---

10 A huge amount of work had been going on since the 1960s based on the assumption that mass is not an intrinsic property of particles—like charge on an electron, for example. With the discovery of the Higgs boson in 2012, the property of mass on particles is explained. Although the discovery of the Higgs boson goes a long way toward solving some of the deep problems in particle physics, there is still more to be done.

comprise. In fact, quarks account for only about 1 percent of the mass of a proton.[11] So it seems that the Higgs particle answers a few important questions, but not all.

To make things more complicated, there is also inertia corresponding to the confined kinetic energy of the quarks, and this contribution to the inertial mass of a proton or neutron falls outside the scope of the Higgs mechanism. So it seems that a combination of effects is responsible for mass, and not just the Higgs field or particle.

In fact, one of the most enigmatic features of quantum physics is that the zero point field, also called the quantum vacuum, is anything but a vacuum. It is more like a plenum, filled with virtual particles that fluctuate into and out of existence. These fluctuations always appear as particle/anti-particle pairs. As we have seen, since they are created spontaneously without a source of energy, these virtual particles momentarily violate the law of conservation of energy, but this is theoretically allowable because the particles annihilate each other within a time limit determined by the Heisenberg uncertainty principle. Apparently it's okay to break a law of nature if you slip it in before Mother Nature has a chance to notice![12]

So, what we have here is an incredible amount of zero point energy, right in front of our eyes. And yet we can't see it. You are being asked to take seriously the existence of a hyper-super-vast

---

11 If 99 percent of the mass of a proton is not to be found in its constituent quarks, then where is it? The answer is that the rest of the proton's mass resides in the energy of the massless gluons—the carriers of the strong nuclear force—that pass between the quarks and bind them together inside the proton.

12 In 2020, researchers at MIT succeeded in measuring the effects of quantum fluctuations at a human scale. In a paper published in *Nature*, the researchers report observing that quantum fluctuations, tiny as they may be, can nonetheless "kick" an object as large as the forty-kilogram mirrors of the National Science Foundation's Laser Interferometer Gravitational-wave Observatory (LIGO), causing them to move by a tiny degree, which the team was able to measure. The boundary between classical and quantum physics continues to shrink.

reservoir of energy that is present everywhere, and strong enough to destroy the earth. And although it's invisible, there is a large amount of experimental proof that—although it may seem crazy—it is true.

Welcome to the wonderful world of quantum mechanics.

Enter Alfonso Rueda

Physicist Alfonso Rueda is one of the world's leading experts in stochastic electrodynamics, an area of physics that concerns itself with the zero point field. After a chance meeting at a colloquium at Lockheed Palo Alto Research laboratories in California (now the Lockheed Martin Solar and Astrophysics Laboratory), he and Bernard began an investigation into the theoretical implications of the ZPF and the law of inertia that was to yield some astounding conclusions.

Early one morning, a few weeks after they first met, Bernard was surprised to receive a lengthy message from Rueda in which he eagerly declared: "I think I've found where inertia comes from. I was able to derive Newton's F=ma." Rueda had been up all night working on the problem. Now, scientists don't usually work to derive an equation like this, because it is generally acknowledged as a "law." They take it as axiomatic. Like 1 + 1 = 2. Naturally, Bernard was curious and impatient to see Rueda's handwritten and equation-rich draft, which he promised to send.

As we have seen, inertia has been a problematic concept in physics. The law of inertia was first formulated by Galileo for horizontal motion and later generalized by Descartes. From the time of Newton into the late 19th century, it has been recognized that inertia is a *sine qua non* (a thing that is absolutely necessary) as a pillar of mechanics. And yet no one had really been able to explain how this essential property came to be. It was simply assumed to be an attribute of matter, a fundamental law of nature.

The best-known attempt to understand the origin of inertia was that of Ernst Mach who, in his 1883 book *The Science of Mechanics*, took an unusual approach that linked the distant matter of stars and galaxies to inertial forces experienced here on earth. This yielded a radically different kind of law, since its causation involved influences exerted over enormous distances in the universe. The only thing that even closely resembles this might be the particle-entanglement theory found in quantum physics today. This perspective is now out of favor, having been superseded by the discovery of either the Higg's field or the ZPF.

Rueda had been hard at work examining the properties of the ZPF—in particular, what occurs under accelerated conditions. And what he found was amazing.

The energy of the ZPF is generally isotropic, meaning simply that it flows in from all directions. But what Rueda found was that, if you get in a rocket and undergo acceleration, the zero point energy as measured by your instruments will show that the energy flow is no longer isotropic. It becomes lopsided. Moreover, the more you accelerate, the stronger and more lopsided the flow becomes. And this is exactly what was needed to explain inertia. If you accelerate in the x-direction, you will experience a resistance coming back at you from the minus x-direction.

What is most intriguing about Rueda's conclusions, however, is that this form of inertia is just good old *electromagnetism*. No strange new fields required.

So what does this mean? We'll have to dig a little deeper into zero point energy and something called "dark energy" to find out.

CHAPTER 7

A New Source of Energy?

Everything we call real is made of things
that cannot be regarded as real.

NIELS BOHR

DARK ENERGY WAS DISCOVERED in 1998 and its discovery shocked astronomers. Edwin Hubble's theory that the universe is expanding had come to be the received wisdom ever since 1920. Astronomers were thus astounded when observations of supernovae revealed that, in addition to cosmic (Hubble) expansion, there is also an anti-gravity effect now known as dark energy.

In physical cosmology and astronomy, dark energy is an unknown form of energy that affects the universe on the largest scales. The first observational evidence for its existence came from measurements of supernovae, which showed that the universe does not expand at a constant rate. Rather, the expansion of the universe is accelerating.[13] The best current measurements indicate

---

13 A central tenet of Einstein's general theory of relativity is that matter tells space how to curve, and space tells matter how to move. But this curved space/time approach is difficult to comprehend. An alternate approach is the polarizable-vacuum (PV) representation, according to which the bending of a light ray near a massive body is due to a spatial variation in the refractive index of the vacuum near the body. The first step in developing Sakharov's conjecture was the work of Puthoff (1989). Expressed in the most rudimentary way, this states that the electric component of the ZPF causes a given nearby charged particle to oscillate. Such oscillations give rise to secondary electromagnetic fields. An adjacent charged particle will thus experience both the ZPF driving forces causing it to oscillate, as well as forces due to the

that dark energy contributes 68 percent of the total energy in the present-day observable universe. This is stunning!

As we have seen, inertia is the property of objects to resist acceleration. An equation that describes this property is F=ma. Force equals mass times acceleration. It is this property that makes solid, stable matter possible.

But this seems to contradict the basic hypothesis presented in this book—that our apparently physical matter-made world is actually a virtual simulation. So how can we talk about a property of inertia that makes the existence of solid matter possible when our entire world is virtual? Needless to say, Bernard was puzzled.

When considering what is real and what is virtual, Bernard had come to accept the existence of an omniscient cosmic consciousness as the creator of the universe and everything in it. An alternative to this would be a non-self-aware field of some sort—something like the ZPF. But it is difficult to picture a non-self-aware field promoting itself to a universal consciousness, so he rejected that possibility, assuming that, as self-aware beings, we superseded such a field. Now Bernard believes in the possibility that the ZPF may, in fact, be an expression of God.

But our model as discussed in the Introduction has a problem when T=0. How do we get back to minus zero or to some other configuration of time as yet unknown? The answer is that God can almost certainly create any form of time that he wishes.

This leaves the existence of a creative consciousness that spins off a vast number of "children" that share in that consciousness as the best possible explanation—that is to say, a pervasive and all-powerful God who populates the universe with living beings

---

secondary fields produced by the ZPF-driven oscillations of the first particle. Similarly, the ZPF-driven oscillations of the second particle will cause their own secondary fields, acting back upon the first particles. The net effect of all this is an attractive force between the particles. The sign of the charge does not matter; it only affects the phasing of the interactions.

and all manner of things for those beings to experience. This model proposes that everything created exists in the mind/consciousness of an omniscient cosmic conscious Creator whose thoughts constitute his actions. And in this model, there is no need for "bricks and mortar"; everything in it is necessarily real; otherwise we would not be here.

Then Rueda discovered that F=ma may actually result from the actions of the ZPF. And this changed everything.

In a research paper published in the *Physical Review* ("Inertia as a ZPF Lorenz force"), Rueda and his colleagues (including Bernard) showed that an oscillator being accelerated would experience a resistance perpendicular to and proportional to the acceleration thanks to the *magnetic fluctuations of the zero point field*. In other words, the law of inertia could be derived within the context of an infinite and pervasive energy that filled every corner of the universe and drove its actions.[14]

This discovery thus has fascinating implications for our model of a universal constructive consciousness—the God model.

The Akashic Field and the Ein Sof

In his intriguing book *Science and the Akashic Field,* Ervin Laszlo draws a comparison between the ancient concept of the Akashic field and the zero point field. The term "Akashic field" is drawn from Theosophy. It refers to a vast database of all thoughts, words, actions, and events that have ever taken place anywhere in the world. The contents of this database, called the Akashic records,

---

14 Arthur C. Clarke was fascinated by this possibility and created a SHARP drive—an acronym for Sakharoff, Haisch, Alfonso Rueda, and Puthoff. He included a SHARP drive in one of his science fiction novels, *3001: The Final Odyssey,* and he and Stephen Baxter even wrote a short story about this concept called *The Wire Continuum* that appeared in *Playboy* magazine in January 1998.

are like the collective unconscious or universal mind proposed by Carl Jung and others.

The very idea of a sea of energy that comprises the collective unconscious has faced a fierce skeptical reaction from scientists. But in the context of an infinite God who seeks to experience a virtual world that his thoughts have created, it makes perfect sense. In short, the Akashic field is simply a means to *access* the Akashic records.

The Akashic field is not the same thing as the zero point field, however. The ZPF is full of particles and random motions. So how does the random ZPF allow for interacting with the Akashic records, which is an organized compendium of all beings, thoughts, and actions in the universe?

In Jewish mysticism, the Ein Sof is the name for the state of God before creation. The term means "infinite" or "without end." In this tradition, the Ein Sof represents God prior to self-manifestation as a divine and endless light. It is the endless One and there is nothing but the Ein Sof. The entire world is thus God in myriad forms and disguises. But the term "Ein Sof" also means "nothing," in the profound sense of "no thing." It exists at the lowest possible energy— zero point energy—but at the same time extends to infinity.

Sound familiar? Could the zero point field possibly be a manifestation of the Ein Sof?

From Zero to Infinity

We have seen that zero point energy has the lowest possible energy density at every point in the universe. For that reason, it is assumed that it is impossible to extract any energy from it. But remember the Casimir cavities we talked about earlier in Chapter 5? These actually allow sub-ground-state energy levels, which opens the possibility for cycling between two energy levels, as in an engine.

But unlike the cycling between heat reservoirs in which the hot reservoir cools over time, the Casimir transition is one of geometry and structure. It does not effectively "cool." So where does the extractable energy come from? Could it possibly have the same origin as the dark energy of astrophysics?

As the universe expands (accelerates), the dark energy in the universe should dilute. But it doesn't. The universe somehow sucks the right amount of energy out of some unknown source to keep its expansion going. A slightly different way of looking at this is that Einstein's cosmological constant is—well, a constant. A third explanation is that the zero point energy *does*, in fact, drain away. It's just that there is so much of it available that it can never be depleted.

So let's look at the Casimir cavity experiment that Haisch and Moddel designed—Harvesting Energy from the Quantum Vacuum (also the ZPF)—in a little more detail. Together they created the Jovion Corporation to carry out this experiment (Jovion Patent 7,379,286, Haisch/Moddel; *www.jovion.com*).[15]

Casimir Cavities

Casimir cavities consist of a space between two parallel conducting plates that are brought into close proximity (less than 1 millionth of a meter). The plates are attracted to one another with a force that increases as the distance between them decreases, and this force will ultimately force the two plates together (see Figure 1).

---

15 Jovion is a clean-technology company working on an approach that could bring affordable energy to homes and places of business from a new, effectively unlimited energy source. It is based on an effect that has been dubbed the Casimir-Lamb shift. Imagine a world that is powered by a device that creates energy on demand. If Jovion is successful, their product—called the **Casimir generator**—will do just that, providing an endless supply of energy that can power automobiles, home appliances, residential and commercial heating and air-conditioning, power plants and factories, the desalination of ocean water, propulsion systems, and many other items.

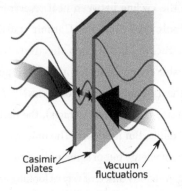

Casimir plates   Vacuum fluctuations

Figure 1. Waves are suppressed inside a Casimir cavity, a phenomenon attributed to electromagnetic energy. Wavelengths longer than the distance between the two plates are excluded from the space between the plates, forming the cavity. This phenomenon, known as the Casimir force, is well known and has been repeatedly demonstrated and measured in a number of credible laboratories.

In the case of an electron, this occurs in less than a billionth of a second. If energy were not constantly being added to replace the energy the electron radiates, matter, as we know it, could not exist. Without a quantum vacuum to keep atomic electrons energized, all atoms in the universe and the universe itself would effectively implode. As long as the atom is exposed to the vacuum, however, the electron energy balance is maintained. If you block access to these frequencies, the electron loses energy and spins off the atom (see Figures 2 and 3).

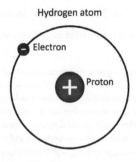

Hydrogen atom

Electron

Proton

Figure 2. A classical orbiting electron.

(2,0,0)

Figure 3. A quantum electron wave function.

How can this phenomenon be turned into a practical device (a Casimir generator)? Jovion has invented a way to extract useful energy from the zero point field (also quantum vacuum field) indirectly by using atoms flowing through Casimir cavities. The cavities required must exclude certain frequencies and must allow the flow of gas through them. These cavities can be made using nanolithography techniques.

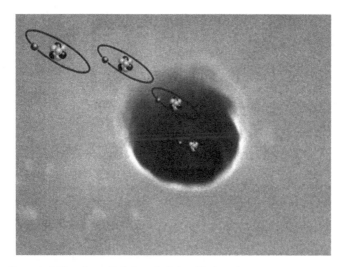

Figure 4. A fifty-micron Casimir cavity formed by laser bursts.

When an electron is energized by the ZPE, a gas atom flows through a single Casimir cavity (see Figure 5). But Jovion plans to use a series of these cavities, converting these photons into usable electricity using photovoltaic solar cells.

The gas flows through Casimir cavities inside a closed-loop system via a small pump, in the same way that freon flows in a refrigerator. Once a gas atom enters the cavity, it releases energy in the form of light as it drops into a lower energy state because it no longer receives energy from the surrounding space. Jovion proposes to harvest this light energy using solar cells (see Figure 6).

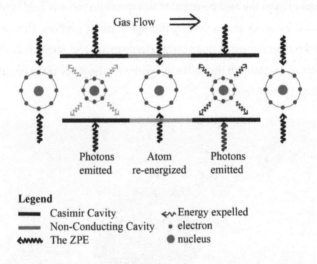

Gas Flow

Photons emitted

Atom re-energized

Photons emitted

Legend

▬ Casimir Cavity

▬ Non-Conducting Cavity

ᔕ∿∿ The ZPE

∿∿ Energy expelled

• electron

● nucleus

Figure 5. Photon emission inside a Casimir tunnel. A gas atom expels photons as it passes through Casimir cavity tunnels. It regains energy upon exiting each cavity, where each electron is energized by the zero point energy. Illustration by Marsha Sims.

When the atom exits the cavity, it is once again exposed to the frequencies that were formerly blocked and the electron is restored to its original energy level. The process is analogous to an automobile radio losing its signal upon entering a tunnel, only to regain it when leaving the tunnel. Gases that may be used in this process are helium, nitrogen, xenon, neon, argon, and krypton. The proposed process is firmly based in classical electrodynamic theories like stochastic electrodynamics.

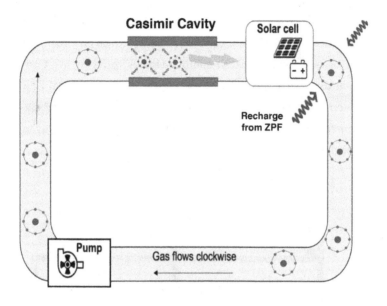

Figure 6. The photons enter a type of solar cell that changes the energy produced by this process into electricity. Here, ZPF waves are denoted by squiggly lines with an arrow at the end. The atoms are re-energized by the ambient ZPF (dark gray squiggly arrow). Illustration by Marsha Sims.

One possible configuration of such a device utilizes stacked disks with Casimir cavity tunnels between each layer (see Figure 7, page 56). In this case, the entire unit would be placed into a sealed cylinder with three openings—one for gas input, one for gas output, and one for wires to funnel electrons from the detector (possibly a photo cell). The device would either block or unblock the zero point energy in a rigorous geometrical fashion.

It is important to note that this process does not violate the second law of thermodynamics, which states that heat flows naturally from an object at a higher temperature to an object at a lower temperature, and never flows in the opposite direction of its own accord. In fact, this process has nothing to do with thermal processes. It relies, not on a temperature differential, but rather on the fact that the Casimir cavity itself has electrical potential. The ZPE level overall reaches a minimum in free space, but in a Casimir cavity, that level may be even lower than the external ZPE level (less than zero), because wavelengths longer than the distance between the two plates are excluded. This allows energy to be drawn out.

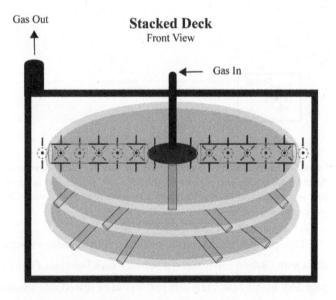

Figure 7. One possible configuration of a working ZPE unit utilizing stacked disks with Casimir cavity tunnels between each layer. The entire unit will be placed into a sealed cylinder with three openings: one for gas input, one for gas output, and one for wires to funnel electrons from the detector (possibly a photo cell). Illustration by Marsha Sims.

The cycle is unending and the energy—well, infinite. The implications for energy production to drive our industries, heat our homes, and light our cities are mind-boggling. Casimir generators would generate electricity locally, and would be sized to match their purpose, depending on the need. For instance, an automobile Casimir generator would have to fit inside an engine space and provide just the right amount of power to run the car. Unlike batteries, however, this generator would be capable of providing a continuous, effectively unending source of energy.

The implications of this process for society are immense. And the implications for the God model as pure consciousness are intriguing.

CHAPTER 8

The Hard Problem of Consciousness

All matter originates and exists only by virtue of a force ... We must assume behind this force the existence of a conscious and intelligent Mind. This Mind is the matrix of all matter.

MAX PLANCK

It is widely believed among cognitive scientists (and among most other scientists as well) that consciousness *must* be something that somehow emerges from complex activities in the brain. Both the "something" and the "somehow" are essentially unknown. Together they constitute what has been called the *hard problem of consciousness,* as stated by philosopher and cognitive scientist David Chalmers:

> Why is it that when our cognitive systems engage in visual and auditory information-processing, we have visual or auditory experience: the quality of deep blue, the sound of middle C . . . It is widely agreed that experience arises from a physical basis, but we have no good explanation of why and how it so arises. Why should physical processing give rise to a rich inner life at all? It seems objectively unreasonable that it should, and yet it does.

And given the amazing success of modern physicalist science in creating technological miracles, each one seemingly out-sparkling the previous, this should perhaps be no surprise.

When Einstein first conceived the general theory of relativity, it was so esoteric that it was said at the time that only two physicists could fully comprehend it: Einstein himself and Arthur Eddington. But today, that little GPS app on your cell phone or in your car pings thirty-one satellites continuously, making precision radio-navigation measurements based on Einstein's theory. We don't need to understand the sophisticated algorithms used to determine our location; we don't even need to know that we are relying on Einstein's theory. We just take the technology for granted. And if GPS were taken off-line tomorrow, millions of us would just have to "wing it."

When confronted with an unsolved problem, savvy scientists naturally assume that it is just a matter of time before that nagging problem will join the ranks of those mysteries and wonders marked "solved." Nobel prizes will be duly awarded and a triumphant sigh of relief will resound through the labs of science and the halls of academia. Indeed, in the light of today's scientific advances, it is difficult *not* to be impressed and to assume that, given time, science will resolve all unanswered questions.

But in his book *From Science to God,* physicist and philosopher Peter Russell observes that this does not seem to be true in the case of consciousness:

> Science has had remarkable success in explaining the structure and functioning of the material world, but when it comes to the inner world of the mind—to our thoughts, feelings and sensations into wishes and dreams—science has very little to say. And when it comes to consciousness

itself, science is curiously silent. There is nothing in physics, chemistry, biology, or any other science that can account for our having an interior world. In a strange way, science would be much happier if there were no such thing as consciousness.

Until now, science has been willing to ignore the critical issue of consciousness, explaining the lively inner world of thought, emotion, and personal identity that we directly experience as the lifeless chemistry in the brain.

We are proposing something different, however. We are proposing that consciousness is *fundamental*. In our view, consciousness cannot ever be explained in terms of something else, because it is itself the fundamental stuff of reality. In fact, we agree with James Jeans, who observed:

> Mind no longer appears to be an accidental intruder into the realm of matter . . . we ought rather hail it as the creator and governor of the realm of matter.

In our model, the only "explanation" required is the direct experience of consciousness that we all possess, and that experience transcends any possible explanation in terms of something more elementary.

So, what is this pervasive consciousness that underlies all reality? Just look at your thoughts. That's it! It's right there, staring you in the face. And what more could you want? Any explanation in terms of other things would only be a step backward. Consciousness is everywhere. It is everything. It is the fundamental stuff of which reality is constructed.

And, believe it or not, this isn't such a new idea.

What Is Consciousness?

Panpsychism is the view that all things have a mind or a mind-like quality. The term, coined by Italian philosopher Francesco Patrizi in the 16th century, derives from the Greek words *pan* (all) and *psyche* (soul or mind). Panpsychism is grounded in the belief that literally every object in the universe—every part of every object and every system of objects—possesses some mind-like quality. In this view, consciousness is an irreducible feature of reality, on par with gravity and electromagnetism.

One of the most resistant problems in science today is the process by which we see and process color. How does the human visual system do whatever it does to yield a perception of a red apple? This is a frustrating problem for materialists, because there is no basis on which to make a comparison. Red, or any other color, can be characterized by individual lines in a spectrum or by a spectral energy distribution, but there is no meaningful way to convert the facts of color into something we can see.

What is the color red "like"? Two healthy, normal observers looking at the same spectrum will agree on the colors. They were taught which color experiences went with which perceptions, so they are in full agreement about the "redness" of apples. *But there is no way short of body switching to guarantee that my red and your red are actually making the same redness impression on our minds.* Are we seeing the same "redness"? The bottom line is that the physical concepts of materialism are insufficient to explain color perception.

Color is, of course, just the tip of the iceberg. Similar logic applies to the perceptions created by the other senses: sound, taste, smell, and touch. And this leaves us with a conundrum. Whatever consciousness does to create our world of things and our daily life experiences must be grounded in more than the usual physical laws, particles, and energies modeled in materialism. There must be some other "thing" involved—something immaterial—that we have not yet discovered.

So it appears that reality has a dual nature. It is both material and immaterial. End of story. Right? It seems not.

The Problem with Dualism

In 2005, astrophysicist Richard Henry published an article in *Nature*, the premiere British science magazine, in which he proposed that *nothing is physical*—the exact opposite of the materialist view—claiming

> The only reality is mind and observations, but observations are not of things. To see the Universe as it really is, we must abandon our tendency to conceptualize observations as things . . . The Universe is entirely mental . . . The Universe is immaterial—mental and spiritual.

While this may seem extreme and incredible, consider the opposite. Consider what speculations about a reality where *nothing is conscious* could mean.

Zombies are purely imaginary creatures indistinguishable from normal human beings, but lacking in conscious experience—like seeing colors. While useful for philosophical speculation, however, the concept of a being without conscious experience is farcical. Nonetheless, some very sober scientists in the fields of physics, philosophy, computation, and artificial intelligence are toying with the idea that human consciousness can be uploaded and stored digitally, creating what might be called digital zombies. One company has even announced that it is in the very early stages of experimenting with this possibility.

And it gets even stranger. Let's say these digital zombies are real. Does that mean that, somewhere in the universe, societies could exist that consist of virtual environments inhabited by virtual residents run by some supercomputer? Would these inhabitants then see themselves and their environment as real, when they are actually an enormous simulation constructed by someone else?

For that matter, are we all "living" our lives on some vast computer that has constructed such a detailed simulation that we are all being fooled? If so, who is doing the simulating? And then there is the question of "digital nesting"—virtual realities that are themselves the creation of yet higher levels of intelligences. In other words, a simulation within a simulation within a simulation. This line of thinking just gets curiouser and curiouser.

This is the net result of assuming that everything that makes up the personality of a human being is nothing more than the composition and organization of the 30 trillion cells that make up the body.

But we all know that personal identity is more than that. And that's where consciousness comes in.

Creative Consciousness

Let's pursue the "mental universe" model proposed by Richard Henry a bit more. If we carry that model to its logical conclusion, we can make the following assumptions:

◊ The entire universe is a single cosmic consciousness. There is nothing else.

◊ This consciousness manifests itself by thinking.

◊ This thinking creates an enormous number of offspring who are of the same nature as the consciousness, but of course infinitesimally "smaller"—like sparks from a bonfire.

◊ This consciousness uses its imagination to create virtual realities that form the worlds these offspring inhabit.

The reason for all this is evident. Through the lives of sentient creatures, the cosmic consciousness is able to experience itself and possibly even to evolve itself.

Put aside your laptop for a moment and imagine instead a Creator who can unwaveringly hold in his thoughts the detailed structure of the entire universe, down to the last atom. Even down to the last subatomic particle. Imagine him able to calculate all the interactions of the particles everywhere in the universe as governed by the laws of nature, which he also created in his thoughts. In this model, all virtual realities are founded on mathematics.

This model could not have been seriously imagined even fifty years ago. But the amazing and rapidly expanding capabilities of today's computers clearly suggest where digital simulations can lead. To arrive at this model, it may be necessary to extrapolate computational capabilities by perhaps 100 orders of magnitude or more. But, as Eugene Wigner warned, it is simply not possible to formulate the laws of quantum mechanics in a fully consistent way "without reference to consciousness." And the advantage of this view is that there is, in the end, only one thing constituting all of reality—*consciousness,* the very thing with which we are all most familiar. Nothing else needs to exist but consciousness, as the source of a realistic but simulated universe.

All that remains is for the creative consciousness to enter into the apparently real life-forms that evolution provides, so that the Creator can experience his creation—everything from human life on planet earth, to the dinosaurs of ancient times, to life in another solar system. This is how this vast consciousness comes to know itself and even to evolve itself—and, dare we say, enjoy or even amuse itself. In this model, as Jeans observed, the universe begins to look "more like a great thought than like a great machine."

In this model, the Creator is the one and only "thing" that is real. Everything else exists only in his consciousness. And that includes us. We are children of the Creator made of the very same consciousness. In that sense, we are *all* God—although of course only to a tiny degree. Because we have free will, however, we deliberately forget that we are sparks of God when we incarnate into a (virtual) physical life on earth.

If there are other civilizations in the universe, the same model holds. There exists one cosmic consciousness whose thoughts create them all, resulting in a universe that is full of interesting "stuff" for us to play with and learn from. And where does all this interesting "stuff" come from? Nowhere! It is all just a virtual reality that exists in the Creator's imagination, organized into computer-like algorithms that run in his mind. Although this doubtless requires unbelievable bandwidth and data storage, it nonetheless solves one of the greatest mysteries of all time.

How did the universe come about? It didn't! It never did and never will exist. It is a non-physical simulation running in the mind of the Creator. Consciousness alone exists.

Stephen Hawking toyed with the notion of *knowing* the mind of God. In this view, we emerge directly *from* the mind of God and are as real as he is. Matter is merely a simulation.

A lovely model. But what evidence is there to support it? We'll explore that in the next chapter.

CHAPTER 9

The Curious Behavior of Photons

The doctrine that the world is made up of objects whose existence is independent of human consciousness turns out to be in conflict with quantum mechanics and with facts established by experiment.

BERNARD D'ESPAGNAT

CONSIDER A VERY SIMPLE OBSERVATION: the reflection of light off a pane of glass. Say that you are in a dark room looking out through a window into a garden on a sunny afternoon. If the glass is clean enough, you won't even see the window, just the garden. Now imagine looking out the same window at night. If there are no lights in the garden, and if you are standing in a well-lit room, what you will see is a fairly faint reflection of yourself and the lit room in which you are standing. That happens because a typical plate of glass reflects only about 4 percent of the incident light off its front surface. This simple observation, known as partial reflection, has profound implications that we'll discuss shortly.

Now consider a stream of water flowing from a faucet. If you turn down the flow slowly, it will finally stop being a stream and become a series of drops. Now imagine doing the same in a properly equipped laboratory with a stream of light from a laser.

Turn it down, down, down until you reach a low enough level that what emerges from the laser is individual photons of light. The human eye is not sensitive enough to see individual photons, but a single photon is easily seen by electronic detectors. Now point the laser at a spot on the surface of a pane of glass and let a detector count the photons that are reflected off the front surface. In any sequence of 100 photons, four photons on average will be reflected. But which photons are reflected and which are not?

The fact is that we can't explain how the photons actually "decide" whether to bounce off the glass or go through it. In fact, the question probably has no meaning. But there's a prior question. What makes some of the photons reflect at all? What is it that is different for those four reflected photons than for the other ninety-six in the sample?

The answer is: There's no difference whatsoever.

This experiment can be done in such a way that a photon is long gone before the next one comes along, so the photons are not conveying information to each other in any understandable sense. And all the photons can be made to hit precisely the same spot on the glass, so it is not a matter of some photons having a different impact point than others. So what tells any given photon that it has the honor of being one of the four reflected?

Nobel laureate Richard Feynman discusses this at length in his book *QED: The Strange Theory of Light and Matter*:

> Try as we might to invent a reasonable theory that can explain how a photon "makes up its mind" whether to go through glass or bounce back, it is impossible to predict which way a given photon will go.

So what does this mean for us? To explore that, we have to look at a field of physics known as quantum electrodynamics—QED.

QED

This curious behavior of photons is one of the starting points of an important field of physics known as quantum electrodynamics. QED studies the interactions of photons and particles at the sub-atomic level and its discoveries have proven to be some of the most stringently tested theories in science. Feynman, one of the founders of the field, called QED "the jewel of physics—our proudest possession." Then he goes on to say:

> I would like to again impress you with the vast range of phenomena that the theory of quantum electrodynamics describes. It's easier to say it backwards: the theory describes all the phenomena of the physical world except the gravitational effect . . . and radioactive phenomena.

In one important QED experiment, it was found that the agreement between the theoretical value of a number and the measurement of that number is so accurate that it is like measuring the distance between New York and San Francisco to the thickness of a human hair.

The incredible precision of QED is very impressive. But how much do we actually understand about it? Even Feynman admitted: "My physics students don't understand it either. That is because I do not understand it. Nobody does."

And that goes for the partial reflection experiment as well. Something has to keep track of how many photons are being reflected and how many are being transmitted. But what? And that something has to have the means and authority to "tell" a photon which of the two possibilities it has to actualize. But what?

To answer these questions, let's take a look at another experiment—the two-slit experiment.

The Two-Slit Experiment

This experiment is easily done in a science lab with a laser and a screen with two narrow slits in it. The laser beam must be wide enough to shine on both slits. When you cover up one of the slits, as the light from the laser passes through the open slit, a pattern will appear on the wall behind it. This pattern is due to the spreading out of light, a process called diffraction—in this case, specifically single-slit diffraction.

Now uncover the second slit. The light going through each slit will still undergo diffraction but, in addition, the light beams traveling through the two slits will interfere with each other. This yields a double-slit interference pattern that is quite different from the single-slit interference pattern. This can be easily explained by picturing light as a wave. With a bit of geometry, you can show how and where the peaks and troughs of the waves will reinforce or cancel each other, yielding the patterns on the wall.

Now assume that the laser beam has been turned down so low that only one photon at a time reaches the two open slits. It is natural to assume that an individual photon can only go through one slit or the other. If that is the case, then, if we let a pattern accumulate on the wall, it should be a single-slit pattern, because each photon can only go through one slit or the other. Right?

Wrong!

As long as both slits are open, a double-slit pattern will build up, even though the photons are passing through one at a time, with each photon long gone before the next one comes along. How is this possible? The conventional explanation is that each photon somehow "knows" that, even though it goes through slit A, slit B is also in the open position, and vice versa.

But how is it possible that a purely material object (the photon) can "know" anything at all? We saw in the first experiment that the different behavior of the 4 percent of photons that were reflected could not be explained by statistical differences in the photons or by random variations in the glass. So something must be "telling" them they should be reflected back instead of being transmitted. Other quantum experiments have led to similar paradoxes. What possible explanation can there be for a single photon "knowing" whether there is one slit open or two?

Hawking and Mlodonow attacked the problem like this:

> In the double-slit experiment Feynman's ideas mean the particles take paths that thread through the first slit, back out though the second slit, and then through the first again; paths that visit the restaurant that serves that great curried shrimp, and then circle Jupiter a few times before heading home; even paths that go across the universe and back. This, in Feynman's view, explains how the particle acquires the information about which slits are open.

It is hard to imagine a more contrived physical explanation. We can think of no way to hardwire the behavior of photons in the glass reflection or the two-slit experiments into a physical law. On the other hand, writing a software algorithm that would yield the desired result is really simple.

In our model, there exists a great consciousness whose mind is the hardware, and whose thoughts are the software creating a virtual universe in which we as beings of consciousness live.

But this thinking leads directly to the inverse problem. If consciousness is the fundamental stuff and it is non-physical, how did the physical universe emerge from consciousness?

That would be quite a feat of creation—making something entirely new and different. Of course, for a religious believer that poses no problem at all. God is credited with creation of the physical universe of matter and energy. From this perspective there are then two components constituting reality: a consciousness which is what it is (whatever that is), and a universe made of matter and energy.

But how would a photon acquire such information? In quantum mechanics, the photon has no definite position from the time it is emitted to when it finally strikes the wall. Discussing this, Hawking and Mlodinow have this to say:

> Feynman's path integral formulation may yield the correct answer mathematically, but it boggles the mind to imagine a more absurd physical explanation. A photon traversing the entire Universe in every conceivable way in zero time! Most physicists are content with getting the correct answer and putting the bizarre Feynman model out of mind as a description of reality. But honestly, if non-scientists were to propose a solution for some phenomenon half as absurd as Feynman's, they would be roundly ridiculed by scoffing scientists. And if you cannot grok the last two paragraphs . . . don't lose any sleep over it.

Perhaps it is time to reconsider the very nature of physical reality.

The Case for Virtual Reality

Could it be that the universe and everything in it is not material stuff governed by rigid physical laws, but rather some kind of virtual reality? This would be consistent with Heisenberg, who saw atoms and elementary particles themselves as potentialities rather than things or facts. And if this is true, does it not follow that consciousness is the only thing that actually exists?

The "crazy" laws of quantum physics that allow atomic particles to do things that violate normal rational physical behavior are the true laws of physics. But we would take this a step further and claim that, in fact, both quantum and classical laws are really the software of consciousness—literally thoughts in the mind of God—and not laws controlling what physical particles do. We'll explore that possibility in the next chapter.

CHAPTER 10

The Software of Consciousness

Before I came here I was confused about this subject. Having
listened to your lecture I am still confused. But on a higher level.

ENRICO FERMI

IN BOTH THE GLASS REFLECTION and the two-slit experiments
of the previous chapter, the problem is what "tells" photons what
to do. Professor of information processing and technology Brian
Whitworth generalizes the problem this way:

> One of the mysteries of our world is how every photon of
> light, every electron and quark, and indeed every point
> of space itself, seems to just "know" what to do at each
> moment. The mystery is that these tiniest parts of the
> universe have no mechanisms or structures by which to
> make such decisions.

There seems to be no way to reduce the behavior of photons in
these experiments to a physical law, or to explain the phenomenon in terms of particles coming in contact with each other. Is it
credible that a photon instantaneously traverses every possible
path through the entire universe, as Feynman posited? And what

kind of bizarre information is possibly being shared between particles in the glass reflection experiment? There must be a simpler explanation.

How about writing a bit of software—an algorithm—that can explain it. The idea of a digital reality whose laws are software was in fact introduced by philosopher and theoretical physicist Nick Bostrom. For example, in the case of the 4-percent rate of reflection, every time a photon is emitted from the laser, let a random number generator select a number between one and 100. Then specify that, if the number turns out to be 25, 50, 75, or 100 (or any other set of four numbers), the photon that triggered the random number will become one of the four reflected.

That does the trick without formulating any complicated laws. And how many other incomprehensible aspects of quantum physics might be understood if we replaced the notion of "laws" with the notion of software algorithms?

We now know that what we perceive as solid, continuous matter is not solid and continuous at all because everything is made up of atoms, which aren't solid at all, but rather insubstantial collections of smaller components. In fact, you can think of them as being *wave functions*. And just what are wave functions? Think of them as the wave versions of particles, just as liquid water and solid ice are two complementary states of water.

In fact, the difference between how we picture reality in classical physics as solid and continuous, and how we picture it in quantum physics is profound. Quantum physics teaches us that, prior to observation, the only "thing" that exists is the insubstantial wave function of particles. In their book *Quantum Enigma*, physicists Rosenblum and Kuttner put it like this:

In quantum theory there is no atom in addition to the wavefunction of the atom. This is so critical that we say it again in other words. The atom's wavefunction and the atom are the same thing; the wavefunction of the atom is a synonym for the atom.

And this insight has profound consequences for how we view the nature of consciousness as well. In fact, one of the biggest and most puzzling scientific questions today is: What is the nature of consciousness and its origin?

Until now, it has been accepted as a given that consciousness somehow arises out of physical matter and processes in the brain. In this view, the physical stuff of particles and energies is real, and consciousness is essentially an illusion, created by the physical. Science has steadfastly refused to take seriously the possibility that it may be the other way around—that consciousness is real and matter is an illusion. This view (consciousness is real), known as idealism in philosophy, has seemed like mumbo-jumbo unworthy of scientific consideration. Samuel Johnson (1709–1784), a literary giant of the 18th century and the second most quoted person in the English language (after Shakespeare), one day was discussing the idealist views of reality proposed by philosopher George Berkeley with his biographer, James Boswell, who recorded the following:

> After we came out of the church, we stood talking for some time together of Bishop Berkeley's ingenious sophistry to prove the nonexistence of matter, and that everything in the universe is merely ideal. I observed, that though we are satisfied his doctrine is not true, it is impossible to refute it. I never shall forget the alacrity

with which Johnson answered, striking his foot with mighty force against a large stone, till he rebounded from it—"I refute it thusly."

But if consciousness cannot be satisfactorily explained as a purely physical phenomenon, and if dualism has not proven to be a viable option, what other alternatives are there? The only logical possibility left is that consciousness is a purely spiritual phenomenon that operates in a universe that is pure thought.

The Universe as a Great Thought

One of the most important lessons we can learn from quantum physics is that we must move beyond the view of the universe as a world of matter, and consider the possibility that the universe is, in the words of Jeans, "a great thought." This view implies not only that our world of matter is not the *only* reality, but rather that our world of matter may, in fact, be a *secondary* component of reality. In the quantum view, reality is not comprised of nuts and bolts and solid stuff, or even of atoms. Rather it is the product of an infinite consciousness that thinks it into being.

Now while that sounds like an intriguing idea, what does it actually mean? After all, the apparently solid world we live in every day certainly feels like stable permanent stuff, not a transient thought. And since the entire universe is over 50 billion light years in size (about a trillion trillion miles) and contains about a billion trillion stars, what kind of thought could possibly contain all that and explain it with any degree of seriousness?

So what exactly does Jeans mean and why should we concern ourselves with opinions that are almost a century old?

In fact, Jeans' view is relevant to evidence that has come to light in the past three or four decades that the universe has special properties conducive to the origin and evolution of life. Astrophysicists now agree that there are enough unexpected "fine-tunings" of physical laws that prove to be favorable for life to arise on the surfaces of planets that some kind of explanation must be found for it. In fact, the universe has proven to be too favorable to the development of life not to address the question, as Bernard did in his two previous books, *The God Theory* and *The Purpose-Guided Universe*:

> The most popular explanation of cosmic fine tuning among scientists is that there must be a vast number of other universes with an array of different laws, the whole set of which has been dubbed the multiverse. Unfortunately, there is not any physical evidence for this, and even the possibility of getting evidence seems extremely unlikely given that the laws of nature would be different in other universes. (If you were a multiverse astronaut, we would predict that you might not fare so well if you managed by chance to cross over into some other universe that had four dimensions of space, two dimensions of time, and five other dimensions that were "none of the above.")

But an equally plausible argument is that the fine-tunings are the deliberate act of some great creative non-physical consciousness seeking experience in physical realms through the life-forms that will inevitably arise. This position is consistent with both the Big Bang theory and with Darwinian evolution.

Both Jeans and Eddington took seriously the view that there is more to reality than the physical universe and more to consciousness than simply brain activity. In his *Science and the Unseen World* (1929), Eddington speculated about a spiritual world and that "consciousness is not wholly, nor even primarily a device for receiving sense impressions." Jeans also speculated on the existence of a universal mind and a non-mechanical reality, writing in his *The Mysterious Universe* (1932):

> Mind no longer appears as an accidental intruder into the realm of matter; we are beginning to suspect that we ought rather to hail it as the creator and governor of the realm of matter ...

So let's postulate that, ultimately, there is one "thing" and nothing else. Nothing except a vast unbounded consciousness (or non-physical mind) beyond space and time—a consciousness that can think, imagine, reason, calculate—and *compute*.

Consciousness as Software

Picture consciousness acting as a self-programming computer, a computer so vast and powerful that it can store and manipulate "bytes of consciousness" and process "software thoughts" sufficient to model the laws of nature and the behavior of every single particle in the universe subject to those laws. In computer terminology, the computational substrate, the platform, is consciousness itself. Consciousness is the hardware. Its thoughts are the software. The scale is beyond imagination—an amazing conscious computer that has no processing or storage limits and thereby can create a simulated reality that models the entire universe. In fact, that simulation would essentially *be* the universe.

In this view, we ourselves are projections of this consciousness into this virtual world. We and all other life-forms interact with this universe using our projected real consciousness to interact with a totally realistic virtual world—even though, in fact, the only thing that is real is the fundamental consciousness along with its projections (us). But to us it seems totally real and we become totally immersed in this virtual world. This virtual world *is* our world.

What is gained by this view? Most important, it obviates the need for anything other than consciousness to exist. The great cosmic consciousness we call God along with his direct offspring are sufficient to explain the universe and all it contains. Real matter becomes a figment of the imagination. It is no longer necessary. All "data" and every "line of code" in this model are thoughts. As long as the simulation is held in the mind of the fundamental consciousness, creation persists and provides us with an arena in which we play out the game of life. And this is how consciousness evolves itself—through the actions and life lessons of its projections—like us.

This view also resolves the origin of space, time, and other universes. Any number of spatial dimensions and time dimensions can be programmed in, as can whatever laws are required to organize and regulate the universe, provided they are consistent. In this model, the cause of the Big Bang is simply the booting up of the simulation. Otherwise inexplicable quantum laws are easily explained. Different universes are simply different subprograms within the fundamental consciousness.

Theoretical physicist Paul Dirac once claimed: "If there is a God, he's a great mathematician." And if this is true, it is no wonder that mathematics is such an effective tool. Reality is digital! The realm in which the pure mathematician works is pure thought: his creations are mental. In this model, the concepts required for our understanding of nature—finite space, expanding space, space of many dimensions, probability as opposed to direct causation—all these concepts and others can be explained as structures of pure thought. If this is so, then, according to Jeans:

> The universe can best be pictured as consisting of pure thought, the thought of what for want of a better word we must describe as a mathematical thinker.

But how is this possible? And what does it mean for the model of consciousness presented here? In the next chapter, we'll explore the logical and philosophical consequences of a universe that consists of pure thought—a virtual universe.

CHAPTER 11

A Virtual Universe

Virtual reality was once the dream of science fiction. But the internet was also once a dream, and so were computers and smartphones. The future is coming.

MARK ZUCKERBERG

TODAY, MANY CREATE AND "PLAY" within virtual worlds thanks to the wonders of computer games and software. But is a virtual *universe* at all a possibility? And is there any evidence for it? Some think there is.

Programming the Universe

In his book *Programming the Universe*, Seth Lloyd contends that the universe itself is one big quantum computer running a cosmic program that produces what we see around us as well as ourselves. Lloyd estimates that it would be possible for programmers to simulate the entire universe in 600 years, provided that computational power increases at the rate established by Moore's law, which states that computational power doubles every two years. While this may be overly optimistic, the dizzying rate at which the computational power of computers is advancing is evident. And whether Lloyd's prediction takes 600 or 6,000 years to come true is irrelevant to our discussion here. What is important here is that it is *theoretically possible.*

Moreover, for Lloyd's virtual universe to manifest, it would not be necessary to simulate the presence and action of every single particle or photon. It's a question of realistic rendering. Take the moon, for example. In Lloyd's model, an object called the moon, along with its principal characteristics, would reside in a database. But, assuming that the simulation would undoubtedly conform to the normal programming practice of minimizing rote computation, it would only be necessary to render the moon to someone who happens to be looking at it, and the rendering would be effective if it had the resolution of, say, a television image (or even less). There would be no need to calculate for the actions of all the atoms and molecules inside the moon. In fact, this kind of rendering is used today in the production of movies and video games.

Even if the rendering were performed for every observer from each person's perspective, that still saves many orders of magnitude of effort over calculating everything all at once. And, yes, this does call to mind the question of whether a tree falling in the forest makes a noise if no one is there to hear it. The difference is that the falling of the tree and the sound of its falling would both be contained in the database. If there was no one around to hear the sound, it simply would not warrant rendering.

In an article in the *MIT Technology Review* (Oct. 10, 2012), Lloyd states:

> The problem with all simulations is that the laws of physics, which appear continuous, have to be superimposed onto a discrete (individually separate and distinct) three-dimensional lattice which advances in steps of time.

But for our purposes, this is good news, because it suggests that there is a difference between continuous and discrete phenomena for which we can test.

A simulated discrete reality has subtle, but essential, differences from a continuous one. And we may be on the verge of detecting these differences. One test of a simulated reality would be that nothing can exist that is smaller than the computational lattice spacing itself. This would predict a pair of features in the spectrum of the most energetic cosmic rays.

But what exactly does this mean? And what are the implications for us?

Measurement and Consciousness

The laws of quantum physics stubbornly link measurement with consciousness and describe behavior—such as entanglement—that can be readily demonstrated, but not explained in terms of hard physical reality. But, as theoretical and mathematical physicist Pascual Jordan notes, in the quantum world "observations not only disturb what is to be measured, they produce it." So how can we resolve these two conflicting views? We've already seen one answer to that question—the QED model discussed in the last chapter.

In the QED model, a software algorithm telling entangled particles, for example, what to do is logically simple and straightforward, because entangled particles far separated in a calculated virtual space have no true spatial separation in the computational realm. Thus the distinction between consciousness and hard physical reality disappears.

Physicist John Bell, who proposed the famous Bell inequality theorem that proved that action at a distance is possible, believed that quantum mechanics revealed that our worldview is incomplete. "A new way of seeing things," he claimed, "will involve an imaginative leap that will astonish us." And indeed, it is becoming increasingly apparent that something beyond ordinary physics is out there waiting to be discovered. In fact, if the universe itself is virtual, then, by definition, it is not something that was created and that simply persists. Instead, it is an ongoing *process*, one that requires continual intention. The software needs to keep running.

Writing in the fifth century, Saint Augustine observed:

> If God's power ever ceased to govern creatures their essences would pass away and all nature would perish. Wherever a builder puts up a house and departs, his work remains in spite of the fact that he is no longer there. But the universe will pass away in the twinkling of an eye if God withdraws his ruling hand.

It is humbling to think that perhaps we have had the answer to this conundrum all along, but have been blinded by science and thus unable to understand it. On the other hand, it is exciting to think that there may be a whole new approach possible to help us unravel the perennial mysteries of what we really are and what this universe is all about.[16]

---

16 For the fearless few willing to suspend disbelief, the adventures in virtual reality of physicist Thomas Campbell as discussed in his monograph *My Big TOE* (Theory of Everything) are recommended reading, and we acknowledge thought-provoking discussions with him and thank him for bringing to our attention this interpretation of reality.

Inside Virtual Reality

Internet-based virtual worlds are being explored everywhere these days. Currently over half a million people have created online versions of themselves—commonly called avatars—to populate these virtual realities. Avatars can have whatever attributes their creators want and can interact with other avatars, participate in individual and group activities, own property, earn money by providing services, and many other actions. In short, they create an artificial reality that becomes more and more real as software and hardware and bandwidth grow. In fact, we are now creating and selling gaming toys that may be primitive analogs of our own reality. One reviewer observed that current virtual-reality programs "blur reality's lines," making illusions feel real. Like it or not, virtual reality is well established in our culture.

Here's a little story to illustrate the point from Bernard's imagination about a virtual-scenario:

> When he clicked the start icon, an ear splitting "chop, chop, chop" roared from out of nowhere and from everywhere at once. It reverberated through his body like a hammer pounding away on his chest. A large army helicopter sent clouds of dust flying every which way as it dropped and touched down no more than 100 feet in front of him. He was pelted by dirt. Jumping from the chopper were half a dozen camouflaged commandos, pouring out with their AK-47s at the ready. And they were racing straight toward him!
>
> One of the commandos bore down directly on him, his AK-47 pointed squarely at him. The tremendous noise and swirling dust were paralyzing. Another second and

the commando would be on top of him. Instinctively, he started to back up. Then the soldier started to run . . . right . . . through . . . him. This threw him off balance and he fell from his chair to the floor.

Jenny, the game master, grabbed Bernard, trying to cushion his fall. "Are you all right?" she asked, helping him to his feet.

"I'm okay," he replied. "Man, that was a lot more realistic than I was expecting, that's for sure."

"Welcome to virtual reality," Jenny said as she lifted off his VR headset goggles.

Of course, Bernard knew all along this was going to be a super-realistic game, but the visuals and the sound were so convincing that he just had an overwhelming reaction—especially when that one commando went plowing right through him and made him fall

"We had your participation mode set to 'ghost avatar,'" Jenny explained, which meant that he would see and hear everything as it occurred, but that, to the soldier avatars, he was invisible—not even there.

"I did not feel anything when that soldier and I collided," Bernard marveled, "other than the chair and the floor. And the surround sound and the high-definition visuals were so intense that I could almost taste and feel being there."

"Taste, touch, and smell," Jenny explained. "Letting you feel a virtual reality for all the senses is still a little tricky, but that will happen before too long. And with an action situation like this, the smell of the chopper and the taste of sand and dirt would fill out the rest of your senses. We have research groups tackling those features. In fact, people doing research in our VR labs are creating scenarios where you can't tell the difference between standing on top of Mount Everest or feeling *convinced* that you are atop Everest. You just experience how realistic we can make the sounds coming from different directions, and the wrap-around field of vision. The next major developments will be suits and boots and gloves that can recreate appropriate sensations that are partly engineering, partly 'cultural,' if I may put it that way."

Is this what awaits us in the future? And if so, why should the concept of a virtual universe seem so improbable to some? Perhaps because it runs smack up against some of our most cherished beliefs—like those of heaven and hell.

CHAPTER 12

The Problem of Heaven and Hell

Scientific progress is measured in units of courage,
not intelligence.

PAUL DIRAC

EVERYBODY WANTS TO GO TO HEAVEN. Even among atheists, there is probably no shortage of those secretly hoping they are wrong and that, if so, they will gain entrance—probably discretely and quietly, through some back door. Perhaps God will let bygones be bygones.

But what could this ultimate destination of heaven be like? The world's religions can't seem to agree on what to expect there. And although most religions claim to tell you what you need to do to get *into* heaven, they actually put more emphasis on what *not to do* to assure that you won't be denied entrance and wind up in the wrong place.

For example, most religions teach that you should not kill except in self-defense, or in defense of another, or in a just war. Some, however, will grant you a direct ticket into heaven for killing the right kind of human being. Some say that, no matter how good you are, you will never be admitted to heaven unless you accept a particular dogma. Some have even claimed that who gets in and who does not is predestined and unchangeable, regardless of how they lived their lives. Welcome, Vlad the Terrible. Sorry, Mother Teresa. Luck of the draw!

As to just *what* heaven is, ideas about that are rather fuzzy. It is generally viewed as a quasi-real place that resembles a happy, peaceful, bliss-filled world inhabited by angels, most of whom are apparently musically inclined and sing in choirs.

But here's a scary thought. Once there, you stay in this heaven forever. You never leave it. And this characteristic of heaven seems to be universal, held across the board by almost all religions. The promise is one of *everlasting* happiness. But that is an oxymoron. Anything you do, no matter how pleasurable at first, can become tedious and eventually even unbearable. And given an infinite amount of time, anything—anything at all—is guaranteed to become an ordeal. The very idea of something going on and on and on, with no end ever possible, promises to be the ultimate nightmare.

No Way Out

Most people who believe in heaven think of it in personal terms. They imagine some glorified, sanctified version of themselves— preferably in their twenties or thirties, in perfect health and good physical shape, without any need for bathroom scales—living in some trouble-free, unpolluted earthlike place. But how would that actually work out?

This point was made in a 1960 episode of the wonderfully creative television program *The Twilight Zone*. The program begins, as usual, with a cigarette-smoking Rod Serling declaiming:

Portrait of a man at work, the only work he's ever done, the only work he knows. His name is Henry Francis Valentine, but he calls himself Rocky, because that's the way his life has been—rocky and perilous and uphill at a dead run all the way. He's tired now, tired of running or wanting, of waiting for the breaks that come to others but never to him, never to Rocky Valentine. A scared, angry little man. He thinks it's all over now, but he's wrong. For Rocky Valentine, it's just the beginning.

Ominous, right?

The story opens with Rocky robbing a pawnshop. He shoots a night watchman and a policeman before he himself is killed by another policeman. He wakes up somewhere, apparently unharmed, in the company of a pleasant character named Pip, who claims to be his guide and who says he is under instructions to grant Rocky's every wish. Naturally, Rocky assumes that he has— miraculously—gone to heaven and that Pip must be his guardian angel. How wonderful.

Life for Rocky goes on in a seemingly endless granting of wishes and good fortune. He plays games and, of course, always comes out the winner. He seeks attention from beautiful women and is never rebuffed. This goes on and on, to the point where Rocky becomes thoroughly satiated and bored. He has had enough of heaven. In desperation, he begs Pip to send him to "the other place." A diabolically bemused Pip replies: "This *is* the other place."

The show ends with Serling's closing commentary:

A scared, angry little man who never got a break. Now he has everything he's ever wanted—and he's going to have to live with it for eternity . . . in the Twilight Zone.

If the thought of endless repetition and knowing that something is never going to stop does not frighten you, you probably have not let the full weight of the word "forever" sink in.

Yet the more human and earthlike we envision heaven to be, the more likely we are to encounter Rocky Valentine's dilemma there. We humans can only have the same things for so long before we want something new and better. The problem is that eternity is a very long time. Indeed, living happily for a true eternity is probably incompatible with human nature. And the idea of heaven being a re-absorption into God does not appeal to the Western mind.

Fire and Brimstone

Eternal flames that never cease and never consume the wretched damned. Third-century Christian writer Tertullian wrote that one of the greatest pleasures of those in heaven will be watching the torments of the damned as they are burned but never consumed for all eternity in the fires of hell. And he was not alone. Seventeenth-century British clergyman Jeremy Taylor, known as a kind of spiritual Shakespeare in his day, wrote:

> Husbands shall see their wives, parents shall see their children tormented before their eyes ... the bodies of the damned shall be crowded together in hell like grapes in a wine press.

Even the great scholar Thomas Aquinas wrote of how "the saints may enjoy their beatitude more thoroughly" if the "sight of the punishment of the damned is granted them." He must have been having a grouchy day when he wrote that. Definitely off message.

But one of the most insane descriptions of hell must be that of 16th-century theologian John Calvin, who believed in the concept of predestination. Calvin taught that, when God creates human beings, he determines—apparently arbitrarily—who will ultimately enjoy eternal salvation in heaven and who will suffer eternal damnation in hell. Even worse, Calvin believed that every newborn infant is born so depraved as to deserve eternal flames forever. But what could be more innocent than a newborn child?

Surely this is a monstrous depiction of God. Only an infinitely cruel God could do such a thing and how could anyone love such a horror? But in fact, therein lies the resolution to the paradox of a God who could be both loving and vengeful.

For it is the dark side of the human imagination that has created God's dark side. God is really nothing like that. Writers and teachers and popes and preachers have just imagined this God and churches have built their dogma around him as a means of social control. Think of the power they gained by teaching that they—and only they—could save you from the eternal flames of hell.

But how do we know that this view of a vengeful God is wrong?

Over the centuries men and women have had mystical transcendent experiences. During those experiences—which generally last for a brief time but, paradoxically, seem to encompass all time—their consciousness melds with the mind of God and that union reveals an intelligence of pure love and bliss, with no thoughts of hellfire and torment. A very different God indeed from the one portrayed throughout the centuries as angry and vengeful.

So let's put emotion aside and try to be purely rational about this. In Gilbert and Sullivan's operetta, *The Mikado*, the emperor sings a humorous song about his dream of creating a system in which the punishment always fits the crime:

> *My object all sublime I shall achieve in time—*
> *To let the punishment fit the crime.*

But an eternal punishment is infinitely disproportionate to any possible crime. A span of 100 years might be within our grasp to understand. A span of 1,000 years pushes the envelope. But try to imagine a million years. And, as long as that is, a million years is still nothing compared to eternity. A million times a billion is still nothing in comparison. Neither is one followed by a billion zeroes. Any number, no matter how large, is effectively zero compared to infinity.

Now think of this in terms of either heaven or hell. In the fullness of time, neither possibility seems attractive. So why would an infinitely intelligent God promise either one? As Mark Twain once quipped:

> I have found it phrased in different ways: Go to Heaven for the climate, Hell for the company. I would choose Heaven for climate, but Hell for companionship.

But is this really the choice that faces us?

Bernard recalls an event from his childhood that prompted him to ask just that question.

The Hot Dog from Hell

At about age eleven, Bernard was having dinner with his best friend and his "sort of Catholic" family. Things were going just fine until he realized that he was chewing a delicious all-meat hot dog on Friday. Visions of hell started appearing in his mind. What a dilemma. What was the rule about swallowing a half-chewed mouthful? Should he rush to the bathroom and spit it out? How he wished that the good Sisters of Providence, who ran his grade school and seemed to know precisely what God wanted, had covered such contingencies. But they never addressed the awkward situation of a forbidden hot dog being eaten as a guest at a friend's house. How could they have missed such a major issue?

Bernard simply had to wing it, hoping that God would either approve or be briefly looking the other way for bigger fish to fry (on Friday!). He could always lie and say that he felt too full to eat the rest of the hot dog, but that would be yet another sin. Of course, such a lie would be only a venial sin, sort of a spiritual misdemeanor, while eating the hot dog was definitely a spiritual felony. Knowing that, if he lived until Saturday afternoon, he could clear the books at confession, he chose the misdemeanor.

But what kind of calculus was this for a young boy to perform? How can any sane person believe that such actions could warrant eternal torture? In retrospect, it sounds like the stuff of a comedy skit, although at the time it seemed very real.

The only logical conclusion is that an eternal hell does not exist. In business terms, you might say the numbers don't support it. In fact, it is an insult to a loving God to believe him capable of such injustice. Should evil acts be punished? Of course. But surely not by eternal damnation at the hands of a supposedly loving God.

Eastern religions may have the right idea. They teach an important concept known as karma. In Christian terminology, this could be interpreted as "reaping what you sow," and negative karma could be interpreted as the "burden of sin." But karmic beliefs are grounded in the possibility of other lives and a belief in reincarnation, not in the inevitability of taking up residence in either an eternal heaven or an eternal hell.

So let's try to clear the air of both the puffy clouds of heaven and the belching smoke of hell once and for all by applying a little simple logic to the problem.

Let's accept the proposition that God is infinitely just. We know that justice requires proportionality between punishment and the offense. And since human life is limited in time, no human can either be infinitely good or commit infinite evil. In fact, both eternal happiness and eternal punishment would be infinitely unjust. And since God is infinitely just, hell cannot exist. But then again, neither can heaven as it is commonly conceived by the religions of the world.

So where does that leave us?

And can science help us answer that question?

CHAPTER 13

Spirituality and Science

*It is true that many scientists are not philosophically minded
and have hitherto shown much skill and ingenuity but
little wisdom.*

MAX BORN

SO IS EVERYTHING WE TALKED ABOUT in the last chapter pure
bunk, as atheists would claim? Hasn't science shown that there is
no God, hence no heaven or hell, or even any other spiritual reality? The official position of the National Academy of Science states:

> Religions and science answer different questions about
> the world. Whether there is a purpose to the universe or a
> purpose for human existence are not questions for science
> ... Science is a way of knowing about the natural world. It
> is limited to explaining the natural world through natural
> causes. Science can say nothing about the supernatural.
> Whether God exists or not is a question about which sci-
> ence is neutral.

This is right and encouraging, in that it recognizes the distinct
spheres in which science and religion operate without declaring
them incompatible.

Still, many people believe that God and science *are* incompatible and that it is the concept of God that should be abandoned. After all, science has proven itself over and over through advances in technology and engineering. But where's the evidence for God? And they may be right, depending on which God they mean to pitch into the dustbin. The eternal tormenter God most definitely needs to be banished.

But we maintain that there is a much better conception of God that can, in fact, be supported by the evidence.

The M-Theory

In fact, the concept of the universe as a great thought discussed in Chapter 11 has acquired a new relevance thanks to recent discoveries about its detailed properties. The fact that the universe seems to be fine-tuned for life to arise and evolve is now a recognized scientific problem in need of a resolution.

We now know that a number of key constants and laws of nature seem to be working together in such a way that the cosmos is full of galaxies, stars, and planets instead of just dilute gas or trillions and trillions of lifeless black holes. Planets in other solar systems are now being discovered that are within a habitable zone from their star—as the earth is from the sun—and are therefore likely places where life can arise. In fact, the latest findings suggest that there may be millions of planets in habitable zones in other solar systems in our Milky Way galaxy.

It turns out that there are about a dozen constants and laws of nature that, if radically different, would have caused stars and planets either not to exist or to have been too short-lived for life to have arisen and evolved. A different strength for the force of gravity, for example, would result in stars that burn out rapidly or never form in the first place, giving rise to black holes instead. Astrophysicist George Ellis, a onetime collaborator with the late Stephen Hawking, concludes:

> What is clear is that life, as we know it, would not be possible if there were very small changes to either physics or the expanding universe that we see around us. There are many aspects of physics, which, if they were different, would prevent any life at all existing . . . We are now realizing that the universe is a very extraordinary place, in the sense that it is fine-tuned so that life will exist.

Of course, this brought an immediate howl of objection from skeptics, who continued to maintain that, of course, our universe is just right for life. If not, we would not be here to take note of that.

This is, of course, correct up to a point. But that's not the end of the story, as mainstream science demonstrates. Mainstream science cannot let it go at that. Instead, they argue that, while it is duly noted that our universe has special properties, they don't like special properties. And so they explain them away by deducing that there have to be lots and lots of other universes with properties different from ours. Thus, although our universe appears to be special, it really isn't. It's all a matter of statistics, they say.

This logic is formally correct, but our (current) best scientific estimate of how many other universes are required to make the statistics work is 10 to the 500th power—in other words, 1 followed by 500 zeroes. This vast number, cited by Hawking and Mlodinow in their book *The Grand Design*, comes out of what is known as the M-theory in physics. But it is a very loose estimate that is sure to change radically as theories evolve. Nonetheless, the estimate will always be a huge number. So, although it's logical, this argument comes with a massive assumption tagging along with it like a ten-billion-pound gorilla. The concept of 10 to the 500th power of unseen and, in principle, even undetectable universes is a lot to swallow, especially since their laws and even their dimensions differ from ours.

Oh, yes, and we almost forgot to mention that there are eleven dimensions accounted for in the M-theory. Try imagining a universe that has four dimensions of length, three different kinds of time, and a couple of dimensions of *je ne sais quoi*. Of course, you can't really picture it, nor can we. In fact, nobody can—not even the most clever M-theorists. To make things worse, there isn't even any unanimity about what the M stands for in M-theory. Wikipedia suggests magic, mystery, or membrane. Some say malarkey.

The Multiverse

This vast multiplication of universes posited by the M-theory goes by the name of the multiverse. Cosmologist and astronomer Martin Rees writes this about the multiverse:

> Our emergence and survival depend on very special "tuning" of the cosmos—a cosmos that may be vaster than the universe that we can actually see.

But if you are tempted to accept the multiverse explanation, keep in mind its drawbacks.

First, it assumes that some kind of quantum fluctuations occurred that created our universe and possibly a huge number of others. But quantum fluctuations require the pre-existence of quantum laws, so where did they come from? No laws, no fluctuations. You cannot get away from the preexistence of *something*, whether that is quantum laws or a conscious Creator. Preexistence of *something* is one of the positions of this book.

The second drawback is that all these other universes that are required to make the statistics come out right can never be detected, because their fundamental laws are different from ours. Their existence is an article of faith among its proponents, since no known proof is possible. Sounds like their objections to the existence of God, doesn't it?

The third drawback is a personal one. If we are nothing but physical beings arising by chance in a random universe, that takes away any ultimate purpose for our lives. It's a bleak outlook that impoverishes us individually and undermines our ethical and moral underpinnings as a civilization.

We propose instead that our universe has the special properties that it does precisely because these properties are the "great thought" that Jeans wrote about, and that great thought has a great purpose behind it.

Purposeless multiverses created by preexisting uncaused quantum fluctuations? Or a purposeful universe created by a preexisting uncaused consciousness? Which is it?

There is a degree of soft evidence on the side of a purposeful universe. Throughout history, people have had transcendent experiences that temporarily pull aside the curtain. And what they see in these visions is a universe of purpose brought into existence and even sustained at every moment by a great conscious intelligence.

Here, we propose that a great conscious intelligence, widely known as God, thought the universe and all its laws into existence.

Randomness or Purpose?

Let us be clear. In the view we propose, Darwinian evolution is essential for fulfilling God's purpose. The unpredictability and novelty afforded by evolution is absolutely necessary; otherwise, existence would be a preordained puppet show.

It is the peculiar character of the universe itself and its origin in the Big Bang that we attribute to an unbounded, perhaps infinite, intelligence, not the micro-engineering of life-forms. It is, in fact, a more impressive feat of intelligence to dream up a few essential laws that can give rise to a universe in which life can evolve, than it would be to tinker around designing creatures like Santa in his workshop.

Choosing between a random or a purposeful universe comes down to a decision on the nature of consciousness. If consciousness is simply a kind of illusion produced by brain chemistry, then the mainstream science view is probably correct. The universe and its inhabitants are just a fluke, a lucky accident, one among many, many other randomly different universes that emerged from pre-existing uncaused quantum laws.

But if the view embodied in Huxley's perennial philosophy is correct, it is the other way around. Consciousness creates reality and the universe is a product of consciousness—that is to say, the great thought enunciated by Jeans. This is a view that mainstream science does not even know how to consider seriously. But, in fact, evidence for a consciously created reality has been lurking in the closet of quantum physics for decades.

The Big Bang

The term "Big Bang" was coined in 1950 by British astronomer Sir Fred Hoyle while he was doing a series of radio lectures on astronomy. Ironically, Hoyle firmly rejected the theory currently denoted by the term, which he intended as a kind of derogatory quip.

However, the Big Bang theory has turned out to be one of the pillars of modern astrophysics and is commonly accepted in the scientific world. It describes the notion that the universe came into being as the result of an unspecified event that caused a massive expansion.

The Big Bang theory is the current prevailing cosmological model explaining the existence of the observable universe. It describes how the universe expanded from an initial state of high density and temperature, and offers a comprehensive explanation for a broad range of observed phenomena, including the abundance of light elements, cosmic microwave background (CMB) radiation, and large-scale structure. The theory holds that, after its initial expansion, the universe cooled sufficiently to allow the formation of subatomic particles and, later, atoms.

But although this prevailing theory is known by the term he coined, Hoyle also suspected that life and indeed the entire universe must be unfolding according to some cosmic plan. "The universe is an 'obvious fix,'" Hoyle said. "There are too many things that look accidental which are not." When asked if he thought some supernatural intelligence was guiding things, Hoyle responded: "That's the way I look on God. It is a fix, but how it's being fixed I don't know."[17]

According to the theory, after the Big Bang occurred, galaxies containing stars were formed. Our own galaxy, the Milky Way, was formed an estimated 13.8 billion years ago. Its evolution began when clouds of gas and dust started collapsing, pushed together by gravity. The cloud contracted under its own gravity and our proto-sun formed in the hot dense center.

It is estimated that our sun was formed 4.6 billion years ago. When its dust cloud collapsed, it formed a solar nebula—a spinning disk of material called the protoplanetary disk. Out of this disk, the planets, moons, asteroids, and other small solar-system

17 As reported by John Horgan, in *Scientific American,* April 7, 2020.

bodies formed. Earth was one of these planetary bodies, sometimes called the "third rock from the sun."

So how did our moon form? This is generally explained by what is known today as the "giant-impact theory," which proposes that the moon was formed during a collision between the earth and another small planet. The debris from this impact collected in an orbit around earth to form the moon, which has been an important subject of scientific research, as well as the source of poetry, music, dance, and romance. We have looked up at it and longed to visit it, despite its inhospitable environment. Finally, in 1969, three courageous astronauts fulfilled an ages-long quest by man and actually set foot on the moon.

As Neil Armstrong told us all in his now famous words, that was a giant step for mankind—a life-changing event that reset our shared perceptions of the universe. Suddenly, we were all part of a world that went beyond our small planet—a part of something greater.

Amazingly, in spite of this event being broadcast to approximately 650 million people, there are those among us who deny it ever happened, as Bernard learned to his surprise some years ago on a sunny April day.

Shooting the Moon

Bernard and his wife, Michelle, were headed for the McDonald Observatory 450 miles away in the west-Texas desert, cruising in their big black Olds 88 Starfire with its 325 horsepower V8 engine. When Michelle began spinning the radio dial, a woman's voice invited them to tune in the next day to hear an interview with the head of NASA's Astronaut Space Medicine Office.

"My friend Darcy claims that moon landing stuff is all fake," Michelle commented.

"You've got to be kidding," Bernard replied. "There's no way anyone could pull that off—not even the CIA. Too many people involved—engineers, scientists. And how could they keep the supposed astronauts from spilling the beans?"

For several minutes, they looked at each other in some annoyance. Then Bernard said: "There were thousands of people out there on the beaches and hundreds of millions watching on TV when that 360-foot-tall rocket lifted off. No way it was faked."

"Darcy claims that there were secret control devices that made sure all the big pieces went down in the Atlantic Ocean, where they sank to the bottom. And the astronauts never left earth. They were held in a maximum-security facility for the duration of the mission. NASA had a super-secret studio where super computers created a totally realistic simulation. Then they snuck the astronauts onto the ship that did the pickup."

"I can't believe this," Bernard said, becoming more and more annoyed with the whole conversation.

"Or maybe they just shot the whole rocket out into space without the astronauts aboard, and then staged the pickup. There are all kinds of ways they could have faked it."

"That is absolutely the most ridiculous thing I ever heard," Bernard bellowed. "Darcy is so gullible! Probably a few too many LSD trips." With that, they dropped the subject. But Bernard couldn't get the idea of a faked moon landing out of his mind.

When they arrived at McDonald, they were treated as honored guests and escorted into a dome where a few astronomers were busy "shooting the moon," performing a lunar laser-ranging experiment that projected a remarkably narrow laser beam all the way to the moon that was reflected back. Laser ranging is a method of measuring the distance between two points with extreme precision. It uses a retro-reflector, a fancy name for a box lined with mirrors that has one opening through which the beam is reflected back to its source. By measuring the time between a pulse of outgoing light and its return, it is possible to determine the distance between the laser and its target—in this case, from the earth to the moon—with an accuracy equivalent to determining the distance between Los Angeles and New York to within 0.25 mm or one-hundredth of an inch.

Bernard turned to Michelle and, with a satisfied smile, pointed out that this experiment alone proved that the Apollo missions did indeed occur, because it relied on retro-reflectors left on the lunar surface by the Apollo astronauts.

This experience showed Bernard how confusion and distrust can result in misinformation, and how lies can block our connection to truth. And it is truth that opens a direct connection to the infinite, divine consciousness. Every step we take in the direction of valuing truth and resisting falsity moves us closer to understanding that God *is* truth.

Synchronicity and Discovery

Today, we take it for granted that signals can be transmitted from one point on earth to another by "bouncing" them off communications satellites. Over 2,000 such satellites are now in orbit. Satellite television providers have millions of customers, and satellite phones allow communication between almost any two points on the planet by relaying satellite-to-satellite-to-ground signals. We even now have satellite WiFi.

But that wasn't always true.

Legend has it that young Isaac Newton was sitting underneath an apple tree one day when an apple fell to the ground, causing him to ask himself what had caused it to fall. And it turns out that this little story is actually true. A manuscript in the archives of the Royal Society of London gives Newton's recollection of the incident, as told to his biographer William Stukeley:

> After dinner, the weather being warm, we went into the garden and drank tea, under the shade of some apple trees . . . he told me, he was just in the same situation, as when formerly, the notion of gravitation came into his mind. It was occasion'd by the fall of an apple, as he sat in a contemplative mood.
>
> Why should that apple always descend perpendicularly to the ground, thought he to himself . . . This in turn led to the insight that the force that brought the apple plummeting down could be the very same force that keeps the moon going around the earth.

And this is a valuable example of how seemingly unconnected synchronous events can sometimes lead to new insights.

Think of the moon falling toward the earth, but also moving forward perpendicular to the fall. The combination of downward falling and forward motion keeps the moon going round and round the earth. The same is true of the earth and the sun. The same force that caused the apple to fall causes the motions of the planets and the moons of the planets.

A deeper and more precise insight into gravity, not as a force but as a curvature of space/time, came along with Einstein's general theory of relativity. But for most applications—including sending astronauts to the moon—Newton's theory of gravity works just fine. A very simple observation led to a great discovery.

In 1960, NASA's Project Echo aimed to put a passive communications satellite into orbit consisting of a metalized balloon acting as a passive reflector of microwaves. The first Echo satellite was lost when the attitude control jets of the Delta launch rocket failed. A second was successfully placed in orbit on August 12, 1960, and, in a subsequent test, a microwave transmission from the Jet Propulsion Laboratory in Pasadena, California, was received in the Bell Telephone Laboratories in Holmdel, New Jersey.

A follow-on Echo 2 satellite was launched on January 25, 1964, but by then the system was already obsolete. Telstar satellites—active rather than passive satellites that relayed and amplified signals via a transponder—began being launched in 1962. This ended the Echo project.

One of the legacies of Project Echo, however, was a large horn antenna. Two Bell Labs scientists, Arno Penzias and Robert Wilson, had ideas for how this super-sensitive twenty-foot antenna could be put to use as a powerful radio telescope. They soon ran into a problem, however. There was an unmistakable noise coming from all directions, day and night, at about the same strength in the microwave range. They assumed, logically enough, that the static they were detecting was coming from the antenna.

This led to the frustrating and painstaking job of ruling out possible sources. It was not radiation from the galaxy, from known extragalactic radio sources, or from the solar system. It was not from above-ground nuclear tests or urban interference from nearby New York City. Even chasing away the pigeons and sweeping out their droppings from the horn antenna had no effect. So they began to look for theoretical explanations.

They finally discovered that the radio noise was coming from the Milky Way region of the sky and that there was a faint electromagnetic radiation present throughout the universe. From this, they were able to conclude that their observations were detecting remnants from a cosmic explosion—the Big Bang that started the evolution of the universe. They shared a Nobel Prize for this discovery.

CHAPTER 14

Matter and Purpose

Anyone not shocked by quantum mechanics
has not yet understood it.

NIELS BOHR

IN HIS BOOK *IS GOD A MATHEMATICIAN?*, astrophysicist Mario Livio recalls giving a lecture at Cornell. As soon as his title slide appeared, he heard a student gasp: "Oh God, I hope not." That appears to be a common sentiment and, for some, even a source of anxiety and heartburn. Anything but math!

But the model of reality that is the basis for Livio's book in fact depends greatly on the likelihood that God is indeed a mathematician. In fact, Livio states: "The universe appears to have been designed by a pure mathematician." And again we hear echoes of the question Einstein asked: "How is it possible that mathematics . . . fits so excellently with the objects of physical reality?"

The great Greek philosopher Plato and his student, Aristotle, had radically different views on the nature of reality. Aristotle took the position that reality was based upon four elements: earth, water, air, and fire. Real stuff. On the other hand, Plato believed that, in addition to the physical world which we all perceive, there also exists a very real world of mathematical forms. The integers 1, 2, 3, etc. exist in this world, as do mathematical concepts like *pi*, the ratio of the circumference of a circle to its diameter, or *c*, the speed of light. The theorems of Euclidian geometry, Newton's laws of motion, and Einstein's laws of relativity also reside there. Plato believed that this ideal realm of eternal, absolute, unchangeable ideas actually existed and was the essence of all things. In this model, the physical world of matter and objects is mere imitation.

Pythagoras and the Pythagoreans had no doubt that mathematics was real, immutable, omnipresent, and more sublime than anything that could conceivably emerge from the feeble human mind. The Pythagoreans literally embedded the universe into mathematics. In fact, to them, God was not a mathematician—mathematics was God.

But what does this actually mean?

The majority of people on our planet believe that some great consciousness—customarily called God—created the universe and everything in it, including us. Creation somehow happened—whether through the Big Bang or in some other way. But just *how* was this accomplished? What made the Big Bang happen? Was God in some way responsible? And if the universe was not created by God, then how can we account for its existence and its persistence?

The Materialist View

In their book *The Grand Design*, Stephen Hawking and Leonard Mlodinow espouse a theory that is based on a view called scientific realism. In a nutshell, scientific realism maintains that the world as perceived by the senses and augmented by scientific instruments is all there is. What you see is what you get. This perspective on reality is also called physicalism, or materialism. In this view of reality, there is nothing but physical matter.

This seems quite reasonable at first. A physical universe made of physical stuff explains it all. But even Hawking and Mlodinow had to admit that, although realism may be a tempting viewpoint, what we now know about modern physics makes it a difficult one to defend and they warned that their theory was "model-dependent," representing the best they could do to explain either the structure or the operation of the universe at that time.

It is perhaps easiest to understand this in terms of an example. At the time of Galileo, in the early years of the 17th century, there were two models of the universe. The Ptolemaic model placed the earth at the center of the universe and had the sun, the five known planets, and all the stars orbiting around it. The Copernican model placed the sun at the center of the universe and had the earth along with the other planets orbiting around it.

While it is tempting to say today that Copernicus was "correct," neither he nor Ptolemy actually had it right. Copernicus was "on the right track," but, compared to our modern view of the universe, he was missing several major elements—like elliptical orbits and orbiting stars. And of course, neither model has anything to say about what the stars actually are—billions of other suns—much less about the (then unknown) galaxies that fill our modern picture of the universe.

In fact, both these conceptualizations are simply models that have required Rube Goldberg-type fiddling and fussing to make them more accurate over time.

When examining the materialist view put forth by Hawking and Mlodinow, we encounter some of the same problems with the model. In fact, we immediately run into a bit of a conceptual problem with this theory. Physical stuff is comprised of particles called atoms. But atoms make up less than 1 percent of the volume of any material, even the densest materials like lead. The rest is mostly empty space. But this "mostly empty space" is certainly in conflict with our everyday perceptions of matter as dense and solid.

Even more problematic is the fact that atoms themselves, the fundamental building blocks of matter, are not even solid all the time. As we have seen, they sometimes act like particles, and sometimes like waves. This wave-particle duality is a major problem for scientific realism, or materialism. For example, the volume of space between the nucleus of a hydrogen atom and its electron is permeated by a wave function that represents only a probability that the point at which the electron is located is a specific location. Thus locating the electron is impossible.

A second problem with this theory is called "entanglement." It is now possible to create pairs of particles—or pairs of photons of light—in a laboratory that rush away from each other at the speed of light, yet remain mysteriously linked together so that a measurement of one affects the properties of the other instantaneously, even if the two are separated by vast distances. And as far as we know, the separation of two entangled particles could extend over our galaxy and beyond. Yet a measurement of the properties of particle 1 would still instantaneously affect particle 2. This is a result of quantum mechanics that has been experimentally verified—although not over the galaxy.

So what is it that entangles one particle with another, and how can this take place instantaneously, when Einstein's special theory of relativity—also tested and verified—specifically forbids instantaneous communication?

Yet another oddity of a purely materialist view of the universe is that particles like electrons can be made to be in two different places at the same time—as can atoms. A *Physics News* item from the *American Institute of Physics* in 2003 carried the headline: "3600 Atoms in Two Places at Once." Physicists Serge Haroche and David

Wineland were awarded Nobel prizes for verifying that this could actually happen. Discoveries like this certainly violate the expectations of realism.

And remember the Heisenberg uncertainty principle. In the real world, if we choose to measure, say, the temperature and humidity of some object, we can do so with (almost) exact precision. But as we have learned, things are very different in the quantum realm. Certain important properties come in pairs—for instance, position and velocity (more precisely momentum). The position and velocity of any object are related in such a way that, the more precisely one is measured, the more the other property becomes uncertain. But this has nothing to do with quality of the measurement. It results from a built-in uncertainty.

The Problem of Purpose

Finally, there is the critical problem that goes to the heart of the materialist view—consciousness. In the real world, if we choose to measure the temperature and humidity of a space, there is every reason to believe that the temperature and humidity already existed before we made the measurement. But things are very different in the quantum realm.

Individual atoms have no temperature or humidity. But they do have other properties. Imagine an atom ejected out of some source and encountering an extremely tiny pair of boxes that are open at the top and the bottom. (This is, of course, a highly idealized setup.) We can choose to make a measurement to determine which box the atom passes through. This would be a particle measurement.

But we could also arrange to measure something different, like the wavelength of the atom. In the previous measurement, we were seeking to know a particle characteristic. Now we seek to know a wave characteristic, which we can obtain by measuring an interference pattern.

So which measurement is the appropriate one to make? The answer is that you are free to choose one or the other . . . but not both. Once you make a particle measurement, you can no longer make a wave measurement of the same atom, and vice versa.

As to what is "really" there, a particle or a wave, recall what Heisenberg had to say:

> The atoms or elementary particles themselves are not real; they form a world of potentialities or possibilities rather than one of things or facts.

And that makes the case for consciousness in a nutshell. It's all a matter of choice.

And choice implies purpose. As Roger Penrose observed:

> The universe has a purpose, it's not somehow just there by chance . . . Some people, we think, take the view that the universe is just there . . . and we happen somehow by accident to find ourselves in this thing. But we don't think that's a very fruitful or helpful way of looking at the universe, we think that there is something much deeper about it.

In the God model, that something is consciousness.

CHAPTER 15

Creation and Consciousness

*If we want to change the world we have to change our thinking
... no problem can be solved from the same consciousness that
created it. We must learn to see the world anew.*

ALBERT EINSTEIN

In *The God Theory* and *The Purpose-Guided Universe*, Bernard argued for three key ideas. The first and foremost is the existence of a great consciousness as the source of our universe. This is in contrast to the prevailing sentiment in the scientific community that the universe arises from some merely random quantum fluctuation. In both the materialist view and the consciousness view, the beginning of the universe took place in a Big Bang event that occurred some 13.8 billion years ago. The question is whether the Big Bang was a random occurrence or an event caused for some purpose.

The second key idea has to do with who we are—whether we are offspring of a great consciousness, each of us sharing a tiny part of that consciousness in a virtual universe, or mere agglomerations of soulless atoms. In Bernard's view, our true nature is not matter, but rather consciousness, the same stuff as the great consciousness of the Creator. Hence we are immortal, because our consciousness is immortal. The physical body is merely what we "put on" in order to experience physical life.

The third idea Bernard considered seeks to answer this fundamental question: If, in fact, God created the universe as a reflection of his infinite consciousness, why did he do it?

Proof of Heaven

Neurosurgeon Eben Alexander's best-selling book *Proof of Heaven* recounts the dramatic experiences of his consciousness traveling through heaven while his body lay comatose for a week in a hospital bed. Does his story prove that heaven actually exists and that he went there, as the title of his book claims? Is this the long-sought hard evidence of a heavenly or hellish afterlife as the title indicates? Did he really encounter God?

Here, we have presented a provocative view of reality—and by reality, we really do mean everything that exists—that does the seemingly impossible task of reconciling the heaven experienced by Alexander with hardcore 21st-century science. That is not to say that we want to "explain away" the heaven that Alexander visited. We have no intention of brushing aside his compelling recollections with glib dismissals or pseudo-explanations. What we propose is a view of reality, and of God, that unites our most advanced scientific knowledge with spiritual evolution. We even suggest that some scientific tests may be available to prove our model in the not-too-distant future.

When *Proof of Heaven* became a best-seller, skeptics fought back. It was all just dreams or hallucinations, they said; nothing worthy of serious scientific consideration, they snorted. But we suggest there is a viable argument to be made against the skeptics. Alexander claimed that his brain was essentially shut down completely; therefore his experiences were clearly not concocted by his brain. Instead, they were actual perceptions gained, not by physical means, but through his abilities as a spiritual being existing independently of his comatose physical body.

Nonsense, the skeptics shot back. All perception is due to brain activity; it cannot be otherwise. Therefore having any perceptions at all is proof positive that Alexander's brain was still functioning at some level, and was the true source of what had to have been hallucinations. No need for God or spiritual mumbo-jumbo to explain that.

But the circular reasoning of the skeptics reveals itself here.

In truth, it all comes down to consciousness. What is it and where does it come from?

Infinite Consciousness

In our model of the universe as pure thought, consciousness is literally *everything*. It doesn't come from anywhere. It is the uncaused cause. It is all there is, and always has been all there is. It is even the cause of space and time, and of the Big Bang. We call this universal consciousness God.

But can an infinite consciousness create anything it wishes to dream up? We actually think not. Consider a square circle. No matter how magically powerful we might be, we are up against a self-contradiction with a square circle. It is absolutely illogical. It cannot even be imagined. The best we can do is picture a square morphing into a circle and vice versa. But that is not the same as simultaneously *being* both a circle and a square. Can't be done, probably not even by God.

What about a skyscraper made out of liquid? This too is impossible, but not in the fundamental sense of a square circle. A liquid skyscraper is impossible because the structural components that you need—beams and girders and walls—cannot be made of something non-rigid like a liquid. But we can "sort of" imagine such a structure—at least if we do not worry too much about inconsistent detail.

But how about imagining something in very realistic detail?

Turn your thoughts to making an imaginary cake. In your mind's eye, picture the ingredients sitting on a counter—a stick of butter, three eggs, a cup of water, and a box of cake mix. Making a cake from these ingredients is simple enough that just about anybody can do it if they follow the directions. So, you pull a make-believe mixer out of a drawer, blend the make-believe ingredients, pour the unreal batter into an unreal pan, and place that into an unreal oven.

Now let's bake another cake—this time, a real cake, one we can actually eat. The hitch is that this second real cake has to be made out of the same imaginary ingredients. But that's a bit of a sticky wicket. There is no way to make a real cake using only your thoughts as ingredients.

The point is this. If consciousness is all that exists, it can no more create a real physical universe out of its thoughts than you can bake a real cake out of make-believe ingredients. But this great God consciousness could *imagine* a universe having any properties he wishes so long as they are self-consistent. Such a virtual universe could be as vast and detailed as necessary to provide the illusion of a physical reality. So why would God go to the bother of creating a real universe—which it may not even be possible to do—when a simulated one is simpler and can be made out of nothing but the thoughts of his consciousness? Who would do such a thing?

We do it—real spiritual beings who are ourselves made of nothing but consciousness and living life in a consciousness-created virtual reality. We are the key that bridges the chasm between spiritual reality and physical reality. In this view, as we saw in chapter 11, the mind of God is the hardware, and the laws of nature and the simulation—comprised of God's thoughts—are the software.

Now what Saint Augustine said 1,600 years ago begins to make sense. The universe is created and continually sustained by the will of God: "Let us therefore believe that God works constantly, so that all created things would perish, if his working were withdrawn."

Consciousness Only

But there is an even simpler view we can take here. Perhaps some transcendent consciousness has created, not a physical reality, but a virtual reality based on its abilities to act like a vast mental computer. At first glance, this may appear to be a trivialization of consciousness as a mere computer. But that is too literal an interpretation.

Think instead of an unbounded intelligence capable of unlimited concentration, able to dream up and keep in mind every detail of an entire universe governed by the laws and logic of that intelligence. And the data storage and computation required for such a Herculean simulation may be only a tiny part of the super abilities of such a consciousness if its potential is unbounded, or even infinite. Creating a simulation of a universe may be just one "project" among many for such an intelligence.

Imagine a consciousness that can unwaveringly hold in its thoughts the detailed structure of the entire universe, down to the last atom, and indeed even down to the last quark and all other subatomic particles. And not only the structure, but the nanosecond-by-nanosecond dynamic behavior of each particle as governed by the laws of nature that this consciousness has also created in its thoughts.

For example, picture the projection of a movie at twenty-four frames per second that we can view with our current technology. Contrast that with the program that runs in the mind of God. Because he is such a super-consciousness, God doesn't even need hard-copy projection. With his enormous intelligence, reality does not have to be projected as literal frames that follow each other. He simply uses his thoughts in place of movie frames.

This hypothesis could not have been seriously imagined even fifty years ago. But the amazing and rapidly expanding capabilities of today's computers clearly suggest where digital simulations can lead. It may be necessary to extrapolate computational capabilities by perhaps 100 orders of magnitude or more, but the advantage of this view is that there is in the end only one thing constituting all of reality—consciousness, the very thing with which we are all most familiar. Nothing else but this consciousness needs to exist to create a realistic but simulated universe.

All that remains is for the creative consciousness to enter into the apparently real life-forms that evolution would provide in order to experience the universe he has created. In fact, this is how a vast consciousness comes to know itself, and perhaps even to evolve itself. As individual projections of consciousness into a virtual universe, we are real; matter is a simulation.

In 1944, Max Planck said:

> As a man who has devoted his whole life to the most clear-headed science, to the study of matter, I can tell you as a result of my research about atoms this much: There is no matter as such. All matter originates and exists only by virtue of a force which brings the particle of an atom to vibration and holds this most minute solar system of the atom together. We must assume behind this force the

existence of a conscious and intelligent spirit. This spirit
is the matrix of all matter.

And this is the heart—and soul—of the God model.

But where does that leave the religions of the world? And how
can we reconcile the model of God as an infinite and pervasive consciousness with traditional Christian interpretations of God as a
"supreme being" whose relationship to us is personal and whose
word is inerrant?

That is the problem we will attack in the next chapter.

CHAPTER 16

The Impossible Inerrant Word of God

From my point of view, God is the light that illuminates the
darkness, even if it does not dissolve it, and a spark of divine
light is within each of us.

POPE FRANCIS

THE SIMPLISTIC IDEA THAT THE BIBLE is the inerrant word of God is not just wrong—it's impossible.

Consider the New Testament. The gospels, epistles, and other material contained there were written in the latter half of the first century, several decades after Jesus's death in about 30 AD. None of the authors ever personally knew Jesus or witnessed his actions while he was alive. The twenty-seven books that comprise the New Testament were gathered together in the second century. Unfortunately, not one original document from that era has ever been found. All we have are copies, and copies of copies, and copies of copies of copies.

In fact, there are over 5,000 hand-copied manuscripts of the New Testament that were produced between approximately the fourth century and the invention of the printing press in 1436. And no two of these hand-copied manuscripts are exactly the same. So, which one of the 5,000-plus documents contains the "inerrant" word of God?

In general, we regard the oldest version of an ancient manuscript as the one more likely to be accurate, because mistakes invariably occur and proliferate when texts are copied by hand. But when considering the authenticity of biblical texts, scholars are faced with two types of problems—inconsistencies between different versions of the same book, and inconsistencies between books. Let's look at a few of these inconsistencies.

Inconsistent "Truths"

The Gospel of Mark presents us with a huge problem. In chapter 16 of the book, Mark gives an account of the resurrection of Jesus, the culmination of Jesus's mission. Verses 1 through 8 tell of three women—Mary of Magdala (or Magdalene), Mary the mother of James, and Salome—going to the tomb and finding the stone at the entrance rolled away. They encounter a young man in a white robe who tells them that Jesus has risen and is not there. He enjoins them to go and tell Peter and the disciples that they will find him in Galilee.

And that's the end of the Gospel of Mark in the oldest version we have today.

But in the New Jerusalem Bible, the resurrection story continues for another twelve verses, in which Jesus appears to others before finally being taken up to heaven. Somewhere over the centuries, a new ending was tacked onto Mark's account.

Now let's look at accounts of the Last Supper. In the gospels of Matthew, Mark, and Luke, the last meal that Jesus has with his disciples—at which he institutes the Eucharist, the rite of communion—is the Passover supper. In Mark's account, Jesus says that a room has been prepared and that the disciples should go there to make preparations . . . which they do. When evening comes, the meal takes place. Afterward, Jesus prays at Gethsemane and is arrested there. He is taken before Pontius Pilate and crucified the next morning. The story is essentially the same in the gospels of Matthew and Luke.

In the Gospel of John, the account starts out the same. Jesus sits down to a Passover meal with the disciples—this is where he washes their feet —and then he is arrested. He is taken before Pontius Pilate the next morning and crucified. The problem arises in verse 28 of chapter 18, where John describes Jesus being led from the house of Caiaphas to the Pretorium—the palace of Pontius Pilate. But now it is morning. The disciples did not go into the Pretorium themselves, John tells us, because they didn't want to be defiled and thus unable to eat the Passover meal. So is John saying that the crucifixion happened *before* the Passover meal?

Another inconsistency arises between the Gospel of John and the other three gospels in the account of the "cleansing of the temple," when Jesus rampages through the area where sacrifices are being sold, declaring that the temple should be a house of prayer, not "a bandit's den." In the gospels of Matthew, Mark, and Luke, this takes place immediately after the so-called triumphant entry into Jerusalem, celebrated today as Palm Sunday, a few days before Jesus was crucified. But in John's account, the cleansing of the temple takes place on some previous visit to Jerusalem early in Jesus's ministry, not at the end. It was after this event, John tells us, that Jesus went into the countryside and began to baptize people (John 2: 13–17). So which account is "inerrant"?

Even the nativity story appears in different versions. In the Gospel of Luke, Joseph and Mary set out from Nazareth to Bethlehem, supposedly to fulfill a decree by Caesar Augustus that a census be performed. Everyone was supposed to return "each to his own town" to be registered. Unfortunately, there is no agreement in any other ancient documents of such a census ever having taken place, although this would be a hugely important event. Moreover, how would you interpret the requirement to return to some ancestral town? Just which ancestral town would that be? Joseph interpreted his place of ancestry to be the town of Bethlehem, because he was of David's lineage. But that was 1,000 years before. There must have been many other towns in which Joseph's ancestors lived at some time in the intervening 1,000 years.

In short, the census story seems to be just a fiction to get Joseph and Mary to Bethlehem so Jesus could be born there. Several other events then take place in Bethlehem, including Jesus's circumcision, key prophesies, and ritual purifications. Then, Joseph and Mary go back to Galilee, to "their own town of Nazareth." But if their "own town" was Nazareth, why did they go to Bethlehem in the first place?

And the account of the nativity given in the Gospel of Matthew is radically different. In fact, there is no mention of any census, nor is there any travel from Nazareth to Bethlehem or any angel appearing to shepherds as in Luke's account. Matthew simply states that, after Jesus was born in Bethlehem, some wise men came to Jerusalem from the East asking about "the infant king of the Jews." This alerts Herod, the current king, that he may have a problem, so he summons the wise men looking for more information. After following a star that exhibits astronomically impossible behavior, the wise men find the house (stable?) where Joseph, Mary, and Jesus are staying.

But let's consider this star. Matthew reports that the star stopped above a house. Now, apart from the fact that the rotation of the earth and not individual stellar motions is the cause of the whole sky seeming to rise and set, and despite stars being suns trillions of miles distant from earth, try finding a specific star that is directly above your house and yet not also above all the other houses in your neighborhood. You can zero in on a house with your GPS, but not, alas, with a maverick star.

In this gospel, the wise men are then warned in a dream not to return to Herod and an angel appears to Joseph in a dream telling him to escape to Egypt with Mary and Jesus. They do as they are told. Meanwhile, Herod slaughters all the male children in the vicinity who are two years old or younger. When Herod dies, he is succeeded by his son, Archelaus, also a no-goodnik. Joseph, Mary, and Jesus return to Israel—but to Nazareth.

Now, some of this may sound like nit-picking. But bear in mind that the New Testament is the most examined book in all of history. And these discrepancies point out why taking the Bible literally, as some do, is problematic. It is no more possible to interpret these conflicting stories as the "inerrant" word of God than it is to assume that 1 + 1 can equal both 2 and 3.

The Problem of Jesus

But all this analysis of the stories of the gospels begs a larger question: Who was Jesus? Or more to the point, what was Jesus? A man? God incarnate? A God/man hybrid?

Most Christians today, ourselves included, believe that Jesus was the son of God, emphasizing his divinity. Some maintain that, if you don't believe in Jesus's godhood, you're not really a Christian. But Jesus also claimed to be the son of man, emphasizing his humanity. To confuse things even more, he also proclaimed that "the Father and I are one," which suggests that Jesus is God himself. So, what is the true nature of Jesus?

There is no indication in the New Testament that Jesus sought to be worshipped as God during his lifetime on earth. The mission and message of Jesus could be interpreted as those of a teacher—or even a prophet—seeking to modify the laws of Moses to make them more humane and compassionate, putting love of one's fellow man above the observance of legalisms and ritual. But nowhere did Jesus say that he was founding a new religion, or that he was God.

But this is exactly what happened after Jesus's death and resurrection. A new religion arose with Jesus deified as one member of a Trinity consisting of Father, Son, and Holy Spirit. And the chief architect of this new theology was not even one of the original twelve apostles who knew and walked with Jesus. The man most responsible for the establishment of what we know as the Christian Church was the great evangelist Paul of Tarsus, a former Pharisee and persecutor of the earliest Christians.

After a life-changing experience in which he was struck by a blinding light and saw a vision of the resurrected Jesus, Paul, also known as Saul, became the most active missionary, taking it on himself to convert vast numbers of non-Jews to the new faith. But his greatest achievement was to define a new religion in which Jesus became the key to mankind's salvation. In modern terms, Paul acted as the chief theoretician and strategist for Christianity. In fact, his letters to various fledgling Christian communities make up a large part of the New Testament.

The mainstream Christian view today is that Jesus was simultaneously a fully human being and fully divine. But is this possible? Of course, you can argue that an all-powerful God can do anything he wants, whether we can comprehend it or not. But as we've seen, that's not entirely true. Even God can't make a square circle. It is a logical impossibility.

And a being that is fully God in a human body is also logically impossible. You cannot put a gallon of water in a two-ounce glass, even if you are God—unless, of course, you cheat by somehow altering the properties of water molecules. Imagine trying to squeeze your human self into the body of an ant. There is simply not enough space, enough bandwidth, enough processing power—call it what you will. A fully human ant is out of the question; so is a fully human amoeba. If we think of God as unlimited and we know that humans are limited physically and mentally, we have to conclude that no limited container can contain something that is unlimited.

The problem lies in the word "fully." But the problem disappears in the model of God as an infinite and persistent consciousness.

Here, we propose that we and all living things are really hosts of God consciousness in virtual bodies. Can a greatly limited bit of God reside in a virtual body? The answer, we believe, is: Of course. A limited amount of God consciousness can fit in a virtual human body, even though "all of God" cannot. And that is precisely what humans and other living entities are—small sparks of an infinite consciousness.

So Why Jesus?

Imagine a king in medieval times. Unfortunately, the peasants in his kingdom are just not behaving as they should and haven't for a long time. They are breaking his laws, constantly fighting with each other and—worst of all—showing no respect whatsoever for him, their lord and master. Year after year, his anger increases, until there comes a time when he has had enough. He is on the brink of teaching them a lesson they will never forget.

But, alas, a part of him still loves them. If only there were some way for them to make up for the offenses and disrespect and insults they have heaped on him. If the peasants would only punish themselves for their behavior, their suffering would soothe and placate the king and eventually cancel out the years of animosity. But there seems to be no sign of that. The peasants are as bad as ever.

So the king hits upon an idea. "I need to see punishment dished out, some pain and blood," he thinks to himself. "Only hard punishment is capable of restoring balance. So, since they will not punish themselves, I will send my beloved son to live among them. And when he claims to be my son, they will do all manner of wicked things to him. I will then accept his torments and blood as theirs and that will placate me and balance will be restored. That will quiet my anger and make me happy."

Does this make the least bit of sense? Would you find satisfaction in the suffering of your own child? Is this imagined king not part lunatic, part psychopath?

Yet this idea of "substitutionary atonement," first enunciated by Saint Anselm in the 11th century, lies at the core of Christian theology. The doctrine claims that the deeds of someone else can erase another's sins and transgressions. Of course, in Christian doctrine, that someone is Jesus and the others are all mankind.

There are several variations of this doctrine, which is perhaps not surprising given the inconsistencies we've already seen in the Bible and the number of theologians past and present who have interpreted it. The version that is most widely held holds that mankind's disobedience to God has generated an enormous debt of guilt that is beyond the capacity of mankind ever to repay. Hence Christ was sacrificed on our behalf and God was thereby placated. Just as in the story of the unruly peasants.

But we don't accept the doctrine of atonement nor do we believe that God is either a lunatic or a psychopath. On the contrary, we believe that he is merciful, benevolent, and loving. Therefore, there must be something seriously wrong with our understanding of God, and of Jesus and his mission on earth.

Theologians have spent quite a bit of time trying to square the circle on the problem of Jesus—2,000 years, in fact. But we argue that no subtle theological reasoning is needed to resolve this problem. God has often been endowed with properties of jealousy, vengeance, wrath, and other charming characteristics. So maybe he is ornery enough to delight in blood-stained appeasement. But we don't believe that the solution lies either in theological esoterica or in an ill-tempered God.

For Christians, there are two fundamental truths concerning Jesus—that he was and still is God, and that he was crucified, died on the cross, and then came back to life after death. And these two essential beliefs are totally consistent with the model of God as infinite consciousness. In our model, there is a steady stream of newly created spiritual offspring of God entering the virtual reality that is the universe, where they are born into bodies to experience life. And the very same process could have brought Jesus to earth as a super-human Godman. Mystery solved.

The evidence given in the gospels for Jesus's resurrection is quite convincing. He is described in several places as appearing to numerous people, some of whom doubted his words and required proof (Matthew 28:17; Acts 1:3; Luke 24:42–43). This tends to support the claim that he really was alive and that there is, in fact, life after death.

But it is important to remember that the power of the story of Jesus lies in his teachings and parables, which were filled with wisdom and emphasized hope, peace, empathy, kindness, love, and patience. Regardless of his theological nature or the inconsistencies in the stories of his life, his words show us how to treat each other with compassion, not to observe religious rules for their own sake. His message that we must do unto others as we would have them do unto us is an imperative for the survival of humanity.

And all of this fits well within the model of God as a creative consciousness.

CHAPTER 17

The God Model

As we near the end of our discussion here, it may be useful to review just what we mean by the God model. Here are a few of its chief characteristics:

◊   There is an unbounded consciousness we call God that exists beyond space and time. This consciousness, which is neither male nor female, created our universe through an event we call the Big Bang, which occurred approximately 13.8 billion years ago.

◊   Consciousness is the only thing that exists. All material things are an illusion, a numerical simulation running in the mind of God. There is no need to *create* reality. It suffices for God to *imagine* a reality and to model its behavior governed by imagined laws.

◊   Wishing to know himself, God gives birth to countless offspring made of the same consciousness as his own.

◊   As sentient beings, we are all immortal incarnations, sparks of the bonfire that is God, and living in simulated reality.

◊ God experiences and evolves himself through the free will actions of his offspring.

◊ We live many lives on earth and elsewhere. We forget these lives when we are born in order to be able to exercise free will in the current life.

◊ There are many other civilizations in the universe, and there are spiritual realms and other universes with properties and dimensions different from our own.

◊ Heaven is our spiritual home, where we reside between lives.

◊ Karma, positive and negative, reflects and determines the course of all our lives.

◊ There is no literal hell. Hell is rather the consequence of our free-will actions, which are perceivable in the afterlife. These may, for a time, be unpleasant.

◊ God's love for us knows no bounds; hence, ultimately, all will be saved after karma is neutralized.

◊ When the light of spirit in the world grows dim, God enters into our world in a direct way as one of us. He appeared to the world as Jesus Christ, the son and incarnation of the creative consciousness we call God.

Of course, this model raises many questions, and we hope we answered some of them here. Yet we know that the model we have presented is complex and raises many other questions as well. So let's explore what some of these might be, and how they are answered by our model.

Questions and Answers

Of course, these answers refer to the model of God as a creative consciousness that we present here and do not claim to come from any kind of divine revelation. They simply attempt to answer fundamental questions in terms of the model.

Why do bad things happen to good people?

The perplexing question of why bad things happen to good people is persistent and heart-breaking. But the God model offers an answer. Remember what we are—sparks of God's consciousness temporarily inhabiting virtual bodies. We have a long road ahead of us; we face numerous lives in numerous circumstances yet to come. Near-death experiences tell us that we can choose to be reunited with lost loved ones in the afterlife if we wish. We can even choose to go through another life experience together.

Admittedly, the promise of a future adventure is scant consolation when we may be suffering a devastating loss here and now. But God will ultimately "make it right" for the simple reason that he is omnipotent and not about to end things on a sour note. Remember, his kingdom knows no end and his glory knows no bounds.

We do, however, have to reckon with karma—the destiny that we earn through our actions and behavior. When we behave kindly, we earn good karma that will result in good things happening to us in the future, although not necessarily in this lifetime. But karma is not a tit-for-tat system. It is much more sophisticated and subtle than that, because there may be other issues that have to be taken into account for our highest good.

At the root of our existence lies free will, without which God would simply be a puppet master who could make anything happen. People, unfortunately, have the free will to commit evil acts that can cause suffering for good people.

Another possibility is that the bad things that happen to us are intended as valuable learning experiences. Or there may be lessons for us to learn about helping others in bad circumstances to strengthen our soul's character. Throughout it all, it's important to remember that God is a loving presence and never abandons us.

Remember, this model is proposed to let the power of God set the stage and then let his powers come to life. God experiences and evolves himself through the free will actions of his offspring.

Is the universe sustained from moment to moment?

Saint Augustine tells us that, if God's power ever ceased to govern creatures, their essences would pass away and all nature would perish. From a scientific perspective, it has been suggested that zero point energy (an expression of God) may play a role in stabilizing atoms and thereby sustaining the universe. This is probably the closest the universe comes to having an on/off switch.

Who or what is God?

There exists an infinite eternal consciousness which is all there is. It fills the universe. It seeks to express itself in all its infinite creative potential. Our reality is one such creation. God is disguised as the world, and the purpose of the game of creation is to uncover the divine, to explore the limits of who we are, to actualize God's self-awareness.

The only thing that exists is consciousness. This consciousness is God and it is what we are made up of. All the material things we see or experience are illusions. It is incorrect, however, to think that a reality consisting of illusions is in any way less real than a reality comprised of material things, because material things do not even exist. Indeed, they *cannot* exist because, as stated above, consciousness is all there is and all that could be.

God/consciousness thinks and this thinking creates an enormous number of offspring who are of the same nature as God, but of course infinitesimally "smaller." God's thoughts manifest as our universe (and possibly others) and its laws. Since God is all there is, all living things on earth and elsewhere are manifestations of himself. In the case of humans, we are the sensible beings by which he is able to know himself. God is a serene, blissful entity.

The entire world is God in myriad forms and disguises. By evolving through space/time, by organizing himself into the complex varieties of existence, God grows and learns endlessly, discovering awareness through each of us—God's countless, inimitable selves.

Did God come from somewhere?

If God came from somewhere, then that place would necessarily be greater than God, or exist prior to him. And the argument could then repeat itself to the next level, and the next, and the next. Astronomers confront the same dilemma with respect to the origin of the universe in a Big Bang. We know that the universe is expanding and that, if we could run the clock backward in time, we would eventually arrive at a massive explosion to which all cosmic matter can, in principle, be traced. The $64,000 question[18] is: How can anything have a beginning when there is no time passing in which it can occur?

---

18 This expression originated in the United States in 1941 on the CBS radio quiz show *Take It or Leave It* where contestants could choose to take a small prize or bet everything on a bigger prize, the highest level being $64,000. It was also the name of an American game show broadcast in primetime on CBS-TV from 1955 to 1958, which became embroiled in the 1950s quiz show scandals.

Perhaps God somehow exists outside time . . . whatever that means. It remains the deepest of mysteries how something can happen when there exists no time in which it can happen.

We propose a model of God that starts with the premise that at first there was nothing. All things began with the logos, which is God. God created the universe with the energy of love and used the Big Bang to boot up a simulation of a virtual universe that appears to us as physical reality.

Is there evidence for the existence of God?

If there can't be a scientific experiment to prove God's existence, how can there be scientific evidence for him? But there can be evidence for something even if we can't test for it. For example, imagine someone asks you: "Do you believe that I love you?" You may believe them or you may not. Either way, your belief does not depend on an experiment that proves the person loves you. Your belief depends on your observations of that person, and those observations can serve as evidence. In the same way, we are able to design an experiment to prove the existence of God that entails numerous observations that point to his existence. We have discussed many of them in this book.

Why is hell impossible?

Hell is claimed to be an everlasting realm of torment reserved for humans who have committed grievous offenses against God. But there is an "open and shut" argument against the existence of such a hell. Even the most evil person can only commit a finite number of sins in one mortal lifetime. But to condemn someone to everlasting torment for a finite number of transgressions is not only unjust, it is infinitely unjust. Yet God is claimed to be infinitely just.

And then there is the absurdity—indeed, obscenity—of taking pleasure in watching the suffering of others. Surely a benevolent God would not do this.

Who or what are we?

We and all living beings are manifestations of God/consciousness. We are identical to the Creator insofar as our essence goes—pure consciousness—but to a far lesser degree. The Creator has spun off a vast number of living beings, each animated by its relatively tiny spark of consciousness. Like the Creator, we and other living beings are immortal—not in our bodies, of course, but in our consciousness. This essence is our soul, or our spirit. Thus, we are eternally living spirits inhabiting virtual mortal bodies.

The reason God has created us in this way is surprisingly simple. He has created an amazing universe and now wishes to experience the wonder of it. We may even be the means by which he evolves himself. He can only do such things by manifesting as a being living in the universe he has created. But for this to succeed, he must forget that he is God. This then opens the door to experiencing the many life experiences each of his spirit children can have. Life on earth certainly offers a vast diversity of life-forms, which is consistent with God's plan for gathering experiences. Consider that there are 8.7 million species of plants and animals on earth.

What is the purpose of life?

Western religions draw a clear distinction between God and humans. God is seen as a loving father or mother, of whom we are the children. God has created humans ideally to enjoy their love of him and, in turn, he reciprocates. This is explicit in the Hindu belief that Brahman (akin to God the Father) is the same as Atman (our spiritual, essential self). Our purpose in life is to experience being and acting in the virtual world created by God. We are the senses of God exploring his creation. Another purpose is to become better beings so that we may rise in any hierarchies that may exist in the afterlife.

Why do we have free will?

God wishes to know himself and experience his creation. Two things are necessary to accomplish this. God must build into his system a forgetfulness, so that his offspring (we) enter life with a clean slate, with no recollection of previous lives and the free will to do what we wish. For some, the weight of previous live(s) might be overwhelming. God, in his justice, provides for this. And after all, humans, good or bad, are really God inside.

Life without free will would be a pointless puppet show.

It may be significant that a growing number of people are remembering their past lives, possibly indicating a shift in human consciousness. Think of how different our civilization would be if we knew and truly believed that we could not outsmart karma. But you can't dodge the karma bullet forever. So think twice before doing something you may regret in a future life.

Can we ask the universe for help?

In astronomy, it is quite clear what constitutes the universe—inanimate stars and galaxies and planets and moons and comets, etc. But these are not wish-fulfilling objects. Needless to say—but Bernard will anyway—he doesn't request miracles from the universe. It would be as silly to him as asking a favor from Ganymede or Callisto, two of the larger moons of Jupiter. So it is useless for us to request miracles from the universe. But we can ask God for assistance, because we are his offspring and he loves us.

New Agers often use the term "universe" when they don't want to invoke the old-fashioned, traditional image of a white-bearded desert-patriarch God sitting on a throne with a staff, floating around on clouds and enjoying being continuously praised in song.

Why do things on earth seem solid?

Everything that's around us seems solid. Think of a whole forest of solid trees. All of these trees are really composed of atoms and we are "tricked" into seeing and believing they are solid when instead science demonstrates that they are just a billion times a billion atoms.

The physical world around us is composed of atoms and these atoms, and eventually molecules, join together to create solid things. Our thoughts commingle with everything in the world around us and cause matter to manifest in our lives. We take this one step further in our model of God as consciousness. We see everything as a simulation, a mind game.

Are there other civilizations in the universe?

From an astronomical perspective, it seems very likely. There are about 100 billion stars in our Milky Way galaxy. That creates a lot of possibilities for exo-solar systems to form. In fact, the opinion of astronomers about life outside earth in the universe has undergone a major shift from negative to positive in the past few decades. As of January 1, 2022, there are 4,905 confirmed exo-planets in 3,629 planetary systems, with 808 systems having more than one planet. As more exo-planets are discovered, the chances for the existence of other civilizations in the universe increase.

UFO sightings offer intriguing evidence to support the idea that there may be other civilizations in the universe. Because the general public has tended to denigrate the term "UFO"— sometimes repeating unverified reports and making fun of possible sightings—researchers have begun to use the term "UAP" (Unidentified Aerial Phenomena) to refer to these phenomena in a deliberate attempt to reestablish their credibility. At the same time, military pilots have reported some amazing encounters with UAPs. In 2021, US intelligence services delivered a groundbreaking report on the credibility of UAPs to Congress.

Some of the Navy's top fighter pilots have encountered mysterious objects that have been tracked by state-of-the-art military radar. Three videos from the targeting pods of US Navy Super Hornets taken in 2004 and 2015 reveal mysterious sphere-shaped objects flying through the sky. Two former Navy pilots described an encounter they had off the coast of San Diego in 2004 on the television program *60 Minutes*, stating that they and two other pilots observed an area of roiling whitewater in the middle of a calm sea, above which small white Tic Tac-shaped objects hovered before disappearing into the ocean.

Incidents like these provide tantalizing testimony of intelligent life coming from elsewhere in the universe. But it is important to keep in mind that there may well be other explanations, some that may be even more mind-boggling to understand than ET. These UAP phenomena have been reported throughout history, even as far back as the Egyptian New Kingdom (circa 1440 BCE) and the Roman Republic (circa 100 BCE). In fact, there is no era that has not reported UAP sightings. If you are interested in these phenomena, we recommend Leslie Kean's book *UFOs: Generals, Pilots and Government Officials Go on the Record*.

What is heaven?

Heaven is a transcendent place where beings like gods, angels, souls, and saints reside, and where we go after leaving our physical bodies. It is our spiritual home, an eternal place to which we go between lives. It is not, as it is purported to be, a paradise with angels flying around with harps on puffy clouds and God sitting on a throne with a staff.

Consider how radically different our modern perspective on heaven must be from that of a medieval peasant. Can we expect to find computers, television sets, automobiles, jet aircraft, telephones, WiFi, cell phones, or miracle drugs in heaven? How about modern services and conveniences like house-cleaning or beauty salons? And what should we expect to find there in the way of careers? Or will we be on permanent vacation? These questions indicate that we seem to imagine a heaven as in some ways similar to the life we experience here on earth.

According to Eben Alexander, who lived through a near-death experience and claimed to visit heaven, heaven is really a much more creative, beautiful place, painted in real time by our personal symbols and imagery. Perhaps it may look more like the heaven depicted in the 1998 movie *What Dreams May Come*, in which the main character paints gorgeous landscapes with his imagination that are very much like his wife's oil paintings.

The heavenly realm may also be a place where we go to consult the Akashic records, review our lives, interface with our personal oversouls or spirits, understand our accomplishments and failures, and plan for how we can improve and grow in an ever-stronger spiritual path in future lives. Also, while in heaven, we can reunite with loved ones who have already made their transitions, meet our own spiritual guides and angelic beings, and discover the essence of our being, which is a spark of God.

Are there other realms outside of our universe?

In chapter 11, we discussed the difficulties of proving a multiverse theory. The theory is difficult to prove since we're not able to visit multiple universes due to our constraints of living in a three-dimensional universe. The multiverse theory can only be proven mathematically. However, according to near-death experiencers, there is the realm of heaven (a kind of spiritual multiverse with many levels of realities) where God, angels, guides, and our souls reside. And this does not need to be proven mathematically.

Does God answer prayers?

God never ignores his beloved children and he's never too busy. He hears every request from us and has all of the resources he needs to answer our prayers. When you are praying, it's important to create a clear intention of exactly what it is that you want. Imagine that you receive your answer using all of your five senses—see it, hear it, touch it, taste it, smell it. If you don't know exactly how you want something to turn out, it's also okay to just pray to God to lead you to your "highest good" and let him work out the details.

Another important aspect of praying is that, if you are in gratitude for God's love, you'll be more open to receiving his gifts. Pay attention to how situations evolve, notice the signs around you, and change accordingly toward your correct path. Sometimes the answers to prayer come in images, sometimes as symbols, and sometimes in Jungian-like synchronicities. Prayers can evolve, so be flexible.

There are several possibilities why prayers may not be answered. God may know that what you pray for may not be for your "highest good." The desired outcome might backfire. Be careful what you wish for, because you might get it. Or your prayer may be in conflict with someone else's highest good. A player on one sports team might pray for his team to win. But you can bet that players on the opposing team are praying for the same thing. So whose prayer is answered in this case?

And remember that we all agreed on plans for our life's path with God and our spiritual guides while we were still in heaven, before being born on earth. God may not answer our prayers if they conflict with these or some vast, eternal plan.

Final Thoughts

It's 9:00 AM AND YOU and far too many other drivers are inching at a snail's pace along the jam-packed freeway. What could be more real than this? Think of the thousands of tons of concrete, the girders of steel, the billboards, the parade of cars. It's as real as real can be—isn't it?

And what are all these things made of?

Atoms, right? Our entire world is made of atoms, 10^{50} or so of them—that is 10 to the fiftieth power; 10 followed by fifty zeroes.

But all of this atomic matter is not what it appears to be. Matter on the atomic scale is actually not solid stuff at all. Viewed from the classical perspective of a point-like electron orbiting a nucleus, an atom is 99.99+ percent empty space. From a modern quantum view, an electron does fill all the empty space in an atom, but it does so as a smeared out "probability thing" called a wave function. And this really is a cloud of probabilities, because it's not possible to predict where an electron will materialize at any given point in time. The laws of quantum mechanics state that an electron is nowhere until it's measured or harvested. In fact, we can now peer into matter deeply enough to image individual atoms.

So, the picture we have in mind of a solid world (and a solid body) is an illusion.[19]

---

19 Anna Blaustein, *Scientific American*, June 28, 2018.

And there is another thing that seems to be unquestionably real: *you yourself.* But not you as a body of atoms. Rather you as the living, thinking conscious being that it *"feels like"* to be a human personality. So what is that made out of?

Seventeenth-century French philosopher René Descartes developed a philosophy that distinguishes radically between mind, the essence of which is thinking, and matter, the essence of which is extension in three dimensions. In this model, known as *dualism,* the nature of the mind is taken to be completely different from that of the body. The opposite of this model is called *monism,* which views all individual souls as the creations of a supreme soul that ultimately merge with the supreme soul after the individual beings die. The differences between these two models are striking.

Monists believe that individual souls are of the same essence as the supreme soul, and that they are ultimately united. Dualists believe they are different and remain separate. Monists believe that individual souls are as divine and powerful as the supreme soul, while dualists see them as powerless. Monists believe that everything in the universe except the supreme creator is illusion, while dualists believe everything in the universe is real. Monists believe that every soul is created from the supreme soul, while dualists believe individual souls are created by some supernatural power other than the supreme soul.

The biggest problem with dualism is lack of any plausible way to explain how some non-physical thing (mind) can interact with some physical thing (body). Descartes clearly identified the mind with consciousness and self-awareness, and distinguished this from the brain as the seat of intelligence. And four centuries after Descartes, we are still at this impasse. On the experimental side, a good deal of research is being done to correlate brain states with consciousness. But how much does it enlighten us to know that, when a specific brain location is stimulated, it is guaranteed to recreate a specific memory?

On the theoretical side, researchers are taking a different approach, asking questions like: What gives water the property of wetness? You can't find wetness in an individual H_2O molecule. Wetness at the molecular level makes no sense. It doesn't exist. In an analogous kind of logic, they speculate that consciousness emerges from the brain, but they don't know how this happens.

Suffice it to say that the origin and nature of consciousness is still up in the air. In our model of God as pure consciousness, we make the assumption that there is literally *nothing* else in existence but consciousness—and this is God. Infinite. Without a beginning. Without an end. Omnipotent—but not quite.

In most religions, God is said to be perfect and all-knowing. There is nothing he needs to do.

But God has in fact done something spectacular! He has created a universe.

So why would God do this? The answer—at least in our model—is that God's ability to know his own splendor requires interaction with some thing or some things that are *not* God. Yet in our model there is *nothing* that is not God. So what's the answer?

God's ability to create is unbounded and his consciousness has no limits. He can bring into existence sentient beings whose key property is that they do not have any memory of being God. This is the secret of the ages. This is the solution to the timeless mystery.

Why do we exist? In order to let God experience his own potential on a clean slate. The objective is experience. What does it feel like to ski down a mountain with twelve inches of fresh powder? God the infinite can ponder this. But God the tiny forgetful spark of consciousness can experience it.

So how does consciousness create a universe? We can see a tentative example of this in the virtual-reality applications available today. Put on a VR headset with the appropriate software and you can fly across mountains on another planet or visit the remains of the Titanic on the ocean floor. Some futuristic computer scientists are contemplating creating entire virtual worlds.

Imagine God thinking about some universe to create. We humans can hold in mind a few thoughts, provided they are not too detailed. Eventually, they fade away. Suppose, however, that God's thoughts are rock steady. Some of these might be thoughts of the laws of nature; some might be thoughts of the little sparks of consciousness that are sent forth to live and thereby acquire experience.

Alan Watts summarizes this scenario in a playful way:

> God also likes to play hide-and-seek, but because there is nothing outside God, he has no one but himself to play with. But he gets over this difficulty by pretending that he is not himself. This is his way of hiding from himself. He pretends that he is you and all the people in the world, all the animals, all the plants, all the rocks, and all the stars. In this way he has strange and wonderful adventures, some of which are terrible and frightening. But these are just like bad dreams, for when he wakes up, they will disappear.
>
> Now when God plays hide-and-seek and pretends that he is you and I, he does so well that it takes him a long time to remember where and how he hid himself. But that's the whole fun of it—just what he wanted to do. He doesn't want to find himself too quickly, for that would spoil the game. That is why it is so difficult for you and me

to find out that we are God in disguise, pretending not to be himself. But when the game has gone on long enough, all of us will wake up, stop pretending, and remember that we are all one single Self—the God who is all that there is and who lives forever and ever.[20]

Watts's playful scenario is totally in keeping with our model of God as pure consciousness.

There exists a single unbounded consciousness we call God who creates all things. He exists beyond space and time. He spins off vast numbers of conscious souls who can interact with each other as his eyes and ears. He creates a universe of enormous potential for sustaining and evolving life-forms. Thanks to his ability to know what is happening everywhere at every level, from the nucleus of an atom to a galaxy of stars, he has no need to make an actual building-block reality. Instead, he makes a world of virtual reality in his consciousness. Material things are an illusion, a numerical simulation running in the mind of God. There is no need to *create* a reality. It suffices for God to *imagine* a reality and to govern its behavior according to imagined laws.

---

20 Alan Watts, *The Book*, p. 15.

Thoughts from Marsha

THROUGHOUT THIS BOOK, we have shared Bernard's thoughts on science, reality, and God. Here, we'd like to share some of Marsha's thoughts on some of these same topics, given in her own words.

Near-Death Experience

When I was sixteen years old, I had a near-death experience. I was visiting some long-time family friends who had a vacation rental on the beach, along with my father, my mother, and my sister. The family had three boys about my age and my sister's, and we five decided to wade out into the ocean and body surf.

We were having a grand time, joyfully jumping in the waves, until a large wave suddenly crashed over me, pulling me under and sweeping me out to sea. Terrified, I watched the shore growing farther and farther away and I screamed to my friends for help. Soon, our parents ran out to see what all the commotion was about.

As I drifted farther out, I gradually went into a more relaxed state, as an alternate reality presented itself to me. The reality seemed to me like a realm between life and death. Several of my ancestors who had died and passed on to the other side appeared around me, holding me afloat in a sumptuous halo of love. They spoke to me mentally, saying: "Don't worry; we will hold you afloat until rescue arrives. It's not time for you to die yet."

I saw my grandmother and grandfather, my great-grandmother, and I think there were more ancestors going back several hundred years, but I don't know their names. It was almost as if a legion of ancestors had rushed to assist me and I was overwhelmed by their love for me. I was in such an emotional state of love that, even though I was a good swimmer, I didn't even need to swim. I just put my head back in the water and floated in a pink halo of light.

After about forty-five minutes, a surfer managed to paddle out to where I was and helped me climb onto his surf board. He took us in a direction perpendicular to the treacherous current and then swung us around in a circular pattern until we reached the beach. Meanwhile, my father had attempted to rescue me by going out in a large inner tube, but he was swept in the opposite direction from where I was. Fortunately, a second surfer was able to rescue him.

I suffered from hypothermia and had to soak in a hot tub for about an hour before I stopped shaking from head to toe from the cold and emotional trauma. As it turned out, the ocean currents were extremely treacherous that day and, unfortunately, another person was swept out to sea and drowned. I'm extremely grateful that I was able to open to my ancestors' love for me and allow them to keep me afloat.

This experience opened me to the other side of the veil of death and allowed me to access memories of past lives. It also blessed me with certain healing abilities that I use when I practice Reiki and Huna.

Marsha's Relationship with God

As a child, before my near-death experience, I loved going to church on Sundays with my family. In church, I felt aglow with the knowledge that God and Jesus Christ loved me. I got this impression from singing in the choir, going to Sunday school, and listening to the reverend's often humorous sermons.

I remember once being in the children's chapel when the reverend joined us to give a special sermon on the glories of God in the universe. He told us that there are so many stars in the universe that they outnumber the grains of sand on every beach and desert on our planet. Bernard and I did some calculations and discovered that, if you assume a grain of sand has an average size and you calculate how many grains are in a teaspoon of sand, then multiply that number by the volume of sand on all the beaches and deserts in the world, the earth has roughly (and we're speaking *very* roughly here) 10^{19} grains of sand—or 7 quintillion, 500 quadrillion grains. There are 10^{11} stars in the Milky Way. Multiply this by at least another 10^9 to include the visible universe and you find that the stars number about 10^{20}. So it seems the reverend was right—but it's close.

I was in awe of the grandeur of the universe that God created and I looked up at the image of Jesus hanging on the wall. I felt as if I'd made a connection with him and tears of joy welled up in my eyes. After my near-death experience, my entire relationship with God and Jesus Christ deepened and became more personalized.

Past Lives

It seems that my near-death experience drew back the veil of forgetfulness that had hidden memories of previous reincarnations of mine. Little by little, I started getting glimpses of past lives that slipped in when I was relaxed. My angel guides started getting my attention with sudden, short loud rings in one ear or another. Right after I heard the ring, guidance in the form of thoughts was presented to me. My guides even introduced themselves to me as a male and female in French Renaissance costumes.

When I was eighteen, I got a bad case of the flu and ran a very high fever. I was home from college and my mom was caring for me. I was asleep in my bedroom when, suddenly, in the middle of the night, I awoke, sat up in bed, and locked myself into the lotus position, although I knew nothing about this position until I was older. I went into a deep trance and found myself going back in time, through many reincarnations and many eras and places around the world. I went back in time so far that I may even have been pre-human.

I saw myself as a female, crouched in a tree in the African savannah nursing a small baby. Next, I saw myself as a wise old Chinese man sitting on the banks of a large river, throwing the yarrow stalks onto the ground to read the I Ching. He looked deeply into the eyes of the future me.

When I awoke from my trance, I remembered that I had traveled through thirty-five past lives, although memories of most of them were quite fuzzy. I remembered some personalities—a monk, a nun, an assistant to a pope, an Egyptian pyramid architect, a terrified on-looker during the crucifixion of Christ, an Italian opera singer, a British banker, and, my most recent incarnation, an Austrian operetta singer who perished in the Holocaust.

These spiritual insights deepened my relationship with God and other holy beings. My point is that, because the veil was pulled open for me, it is easier for me to let God into my awareness.

The Path to Peace

In 1998, I had a major awakening of the heart at a Gangaji Satsang or Buddhist gathering, where I learned about self-inquiry as a path to experiencing the truth. I was sitting on a pillow meditating when the spiritual leader of the retreat came into the room, looked around, and then looked straight into my eyes. It was as if I'd been hit by a nuclear explosion of lights and love. I closed my eyes and saw swirling colors—purples, magentas, and blues—coming right out of the top of my head.

After bathing myself in these colors for a while, I opened my eyes and looked around the room as tears of joy welled up in my eyes and my heart pulsed. The entire room was bathed in a brilliant white light and I could see fingers of white light connecting each person in the room. It seemed as if we were a community of God's children. Later, I discovered I had had what is called an orgasm of the heart, in which I was overjoyed about establishing a very deep connection with God. It was like what the New Testament describes as "the peace that passes all understanding."

My point is that we all are citizens of God's universe and he loves each one of us very dearly. He is able to fill us with his love whenever we can quiet our minds. We can all do this if we take the time to stop our "monkey minds," which hop around in a circle like a monkey with endless thoughts. God speaks to us in a still, small voice when we're able to listen, but our minds must be quiet.

Since then, I have sometimes had difficulty getting to a deep level of meditation like that. Currently, my method is to be in a quiet place, close my eyes, and count backward from 100 to 1 as many times as needed to get into a meditative trance. Sometimes my thoughts come in so quickly that I have to begin counting backward faster than they appear. Gradually, however, my thoughts slow down and my mind relaxes and I start to see colors. One of the clues I get that I'm really deep is when I see a three-dimensional lime-green circle surrounded by a magenta or purple halo. That is when I can start knowing that the thoughts in my mind are messages from God, my guides, and other angelic beings.

It's also important to be in a state of gratitude, because this opens your mind for optimism, love, and God's still small voice. The good news is that we can all achieve this if we only take the time to still our monkey minds. It may take a lot of practice, but it's worth it. If everyone made the effort to do this—to dialogue directly with God on a daily basis—our world would be catapulted to a higher level of awareness and the human race would become ever more enlightened!

Connecting to God Through Music

I experience and connect to God through music—the universal language that crosses all barriers and connects humanity on a higher level of awareness. As a professional music teacher, I know that parents are eager for their children to learn music because they innately understand the importance and power of music in our society. Teaching allows me to share the magic of music with my students and watch them grow as musicians. It is also very therapeutic for me. It lifts my mood and energizes me.

One of my students recently told me that playing the piano helps her to concentrate, because, in order to succeed, she has to block out all other thoughts. Music thus becomes a form of meditation for her and connects her to God. Everyday thoughts fade once she becomes immersed in the beautiful stream of music that is flowing through her.

I've also used music as a healing tool for physically handicapped students. Research shows that music is therapeutic for various mental health conditions, including depression, trauma, autism, and schizophrenia. Music can be a medium for processing emotions, trauma, and grief. It can also be used to regulate or calm patients who suffer from anxiety or other problems. It can help people of all ages with learning disabilities, brain injuries, and physical disabilities.

Math is also essential in music production, including recording and mixing. The digital audio workstation (DAW) software used by producers relies on mathematical algorithms for signal processing, compression, and equalization. Moreover, electronic music genres such as techno and Electronic Dance Music (EDM) are often created using mathematical concepts such as algorithms and fractals.

The curriculum of the medieval university consisted of the trivium (grammar, logic, and rhetoric) and the quadrivium (arithmetic, astronomy, music, and geometry). Together they were thought to lead students to see a "unified idea of reality," which shows the close connection between mathematics, music, and astronomy.

Singing in church was exhilarating for me and I also got a very powerful connection to God when directing the choir or leading the congregation in song. The connection I felt to God's love and beauty was magnified as other voices joined in. An elaborate musical church service can be interpreted as a gift from us to God, which in turn opens a path from God to us. Other musicians have told me that they feel energized, pumped up, and electrified when performing. They report feeling present and focused on the joy of the moment.

I feel exhilarated when I sing in community musical and operatic theater for the joy of it. Whenever I sing an operatic role in front of a live audience supported by an orchestra, I become a vessel for the spirit of God. He flows into my mind, through my vocal cords, and out of my mouth. I feel a concordance with others, as if we all become one with God and with each other. It is simply magical! During these performances, I concentrate on the drama and simply become a vessel for the flowing music, for life, and for God.

Upon reflection, I now see that these experiences are similar to the one I had with Gangaji and my awakening of the heart. Music opens my heart and everyone else's. That is why I've dedicated a large part of my life to it.

The Harmony of the Spheres

In medieval times, there was a close association of music with mathematics and astronomy, and music was thought to lead to a "unified idea of reality." This idea was developed by thinkers like astronomer Johannes Kepler and philosopher Francesco Giorgi, whose treatise *De Harmonia Mundi* argued for a totally numerical model of the universe grounded in the harmonics of music.

In fact, music is closely tied to mathematics through harmonics. Remember James Jeans's assertion that the universe appears to have been designed by a pure mathematician? Well, Johann Sebastian Bach, the father of modern music, actually managed to devise a system of tuning that is based on mathematics. In this system, the twelve notes of the octave of the standard keyboard are tuned in such a way that it is possible to play music in all major or minor keys without sounding perceptibly out of tune because each step is the twelfth root of two ($^{12}\sqrt{2}$, $2^{1/12}$, or 1.0594631).

One of the most important organizing principles in this system is called "the circle of fifths," which lays out the twelve chromatic pitches in a sequence of perfect fifths, usually depicted as a circle (see Figure 8). When C is chosen as a starting point, the sequence is C, G, D, A, E, B (=C♭), F# (=G♭), C# (=D♭), A#, E#, B#, F. Continuing the pattern from F returns the sequence to its starting point of C. This order places the most closely related key signatures adjacent to one another.

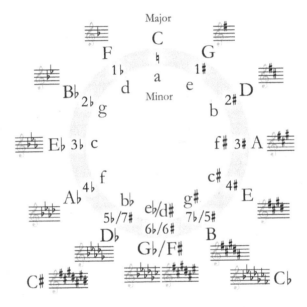

Figure 8. The circle of fifths.

And other aspects of music are quite mathematical as well. Rhythmic relationships are portrayed as fractions, as are time signatures—notational conventions used to specify how many beats (pulses) are contained in each measure (bar), and which note value is equivalent to a beat. Time signatures have an upper and lower number and are portrayed as fractions like 4/4, 3/4, or 6/8. These numbers are not the same as regular fractions, however.

For example, in 3/4 time, the upper number (3) represents how many beats in a measure, while the lower number (4) represents the type of rhythmic note that gets one beat. (Here it is the quarter note.)

Figure 9. Three-quarter (¾) time signature.

Rhythm refers to a variety of note and rest durations that appear in the context of the time signature, or meter. Notes can last for any length of time—for an entire beat or even multiple beats. Notes can be shorter than a beat as well. Typical rhythmic notes are whole notes, half notes, quarter notes, eighth notes, and sixteenth notes. More complicated rhythms can be achieved by adding half of a note's value to itself, in which case the note is said to be "dotted." For example, a dotted half note 𝅗𝅥. receives two beats plus an additional beat (half of itself), for a total of three beats.

Rhythmic Notation for Notes and Rests

Name	Note	Rest	Beats	1 $\frac{4}{4}$ measure
Whole	𝅝	▬	4	𝅝
Half	𝅗𝅥	▬	2	·𝅗𝅥 𝅗𝅥
Quarter	𝅘𝅥	𝄽	1	𝅘𝅥 𝅘𝅥 𝅘𝅥 𝅘𝅥
Eighth	𝅘𝅥𝅮	𝄾	½	𝅘𝅥𝅮𝅘𝅥𝅮 𝅘𝅥𝅮𝅘𝅥𝅮 𝅘𝅥𝅮𝅘𝅥𝅮 𝅘𝅥𝅮𝅘𝅥𝅮
Sixteenth	𝅘𝅥𝅯	𝄿	¼	𝅘𝅥𝅯𝅘𝅥𝅯𝅘𝅥𝅯𝅘𝅥𝅯 𝅘𝅥𝅯𝅘𝅥𝅯𝅘𝅥𝅯𝅘𝅥𝅯 𝅘𝅥𝅯𝅘𝅥𝅯𝅘𝅥𝅯𝅘𝅥𝅯 𝅘𝅥𝅯𝅘𝅥𝅯𝅘𝅥𝅯𝅘𝅥𝅯

Figure 10. Notes, rests, beats, and how they compare to each other mathematically.

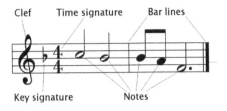

Figure 11. An example of musical notation with a treble clef, a key signature (Bb for F major), a time signature of 4/4, bar lines, half notes, eighth notes, and a dotted half note. The first measure (left) has two half notes (four beats); the second measure (right) has two eighth notes (one beat) plus a dotted half note (two beats), for a total of three beats.

The mathematical nature of musical construction and notation makes me wonder whether God uses music as a mathematical expression through sound. Or is it possible that music is mathematical because the universe was designed by a pure mathematician? Maybe it's both?

And this may be why musicians and non-musicians alike seem to experience God through music. Music has the greatest influence on us of anything abstract in the world. It under-girds all of society. It is impossible to get away from it, because it is pervasive around the planet, even though the forms it takes vary tremendously—from Muslim calls to prayer, to classical symphonies, to operas and Masses, to rock and pop and rap. One hymn even claims that: "All nature sings and round me rings the music of the spheres." Perhaps music, by its mathematical nature, is a universal expression for civilization and God.

Bibliography

Alexander, Eben. *Proof of Heaven: A Neursosurgeon's Journey into the Afterlife.* Simon & Schuster, 2012.

Assante, Julia. *The Last Frontier, Exploring the Afterlife and Transforming Our Fear of Death.* New World Library, 2012.

Baldwin, Neil. *Edison: Inventing the Century.* University of Chicago Press, 2001.

Barbour, Julian. *The End of Time: The Next Revolution in Physics.* Oxford University Press, 1999.

Beane, Silas R., Zohreh Davoudi, and Martin Savage. "Constraints on the Universe as Numerical Simulation." Cornell University Library, arXiv: 1210.1847v2 [hep-ph], 2012.

———. "The Measurement That Would Reveal the Universe as a Computer Simulation." *MIT Technology Review*, Oct. 10, 2012.

Bostrum, Nick. "Are You Living in a Computer Simulation?" *Philosophical Quarterly*, Vol. 53, No. 211, pp. 243–255, 2003.

Campbell, Thomas. *My Big TOE (Theory of Everything).* Lightning Strike Books, 2003.

Chopra, Deepak. *The Path to Love.* Harmony Books, 1997.

Chopra, Deepak and Leonard Mlodinow. *War of the Worldviews: Where Science and Spirituality Meet—and Do Not.* Three Rivers Press, 2011.

Ciufolini, Ignazio and John Wheeler. *Gravitation and Inertia.* Princeton University Press, 1995.

Danison, Nanci. *Answers from the Afterlife.* A. P. Lee & Co., 2016.

———. *Backwards: Returning to Our Source for Answers.* A. P. Lee & Co., 2007.

Dossey, Larry. *One Mind: How Our Individual Mind Is Part of a Greater Consciousness and Why It Matters.* Hay House, 2013.

Eddington, Sir Arthur Stanley. *Science and the Unseen World.* The Macmillan Company, 1929.

Ehrman, Bart. *Jesus Before the Gospels: How the Earliest Christians Remembered, Changed and Invented Their Stories of the Savior.* HarperOne, 2016.

——— *Did Jesus Exist? The Historical Argument for Jesus of Nazareth,* HarperOne, 2013.

Feynman, Richard. *QED: The Strange Theory of Light and Matter.* Princeton University Press, 1985.

Gangaji. *You Are That!* Self-Published, 1996.

Gribbin, John and Martin Rees. *Cosmic Coincidences: Dark Matter, Mankind, and Anthropic Cosmology.* Bantam Books, 1989.

Guillen, Michael. *Can a Smart Person Believe in God?* Thomas Nelson Inc., 2006.

Haisch, Bernard. *The Purpose-Guided Universe: Believing in Einstein, Darwin and God.* New Page Books, 2010.

———. *The God Theory: Universes, Zero-Point Fields and What's Behind it All.* Weiser Books, 2006.

Hawking, Stephen. *Brief Answers to the Big Questions.* Hodder & Stoughton, 2018.

Hawking, Stephen and Leonard Mlodinow. *The Grand Design.* Bantam Books, 2010.

Heisenberg, Werner. *Physics and Philosophy.* Harper Perennial Modern Classics, 2007.

Hitchens, Christopher. *God Is Not Great: How Religion Poisons Everything.* Warner Books, 2007.

Hoffman, Donald. *The Case Against Reality: Why Evolution Hid the Truth from Our Eyes.* W. W. Norton & Company, Inc., 2019.

Huxley, Aldous. *The Perennial Philosophy.* Harper & Row, 1944.

Jahn, Robert and Brenda Dunne. *Margins of Reality: The Role of Consciousness in the Physical World.* Harvest Books, 1987.

Jammer, Max. *Concepts of Mass in Contemporary Physics and Philosophy.* Princeton University Press, 2000.

Jeans, Sir James. *The Mysterious Universe,* New Revised Edition. Macmillan Company, 1932.

Kastrup, Bernardo. *Dreaming Up Reality: Diving into Mind to Uncover the Astonishing Hidden Tale of Nature.* iff Books, 2019.

————. *The Idea of the World: A Multi-disciplinary Argument for the Mental Nature of Reality.* iff Books, 2019.

Katra, Jane and Russell Targ. *The Heart of the Mind.* New Work Library, 1999.

Kean, Leslie. *UFOs: Generals, Pilots and Government Officials Go on the Record.* Crown Archetype, 2010.

Lederman, Leon. *The God Particle: If the Universe Is the Answer, What Is the Question?* Delta, 1993.

Livio, Mario. *Is God a Mathematician?* Simon & Schuster, 2009.

Lloyd, Seth. *Programming the Universe: A Quantum Computer Scientist Takes on the Cosmos.* Vintage Reprint Edition, 2007.

Long, Jeffrey, MD and Paul Perry. *God and the Afterlife: The Groundbreaking Evidence for God and Near-Death Experience.* HarperOne, 2016.

Long, Max Freedom. *The Secret Science Behind Miracles.* DeVorss & Co., 1948.

Loudon, Rodney. *The Quantum Theory of Light,* 2nd Ed. Oxford University Press, 1983.

MacGregor, Malcolm. *The Enigmatic Electron.* Kluwer Academic Publishers, 1992.

Matt, Daniel. *God and the Big Bang: Discovering Harmony Between Science and Spirituality.* Jewish Lights Publishing, 1996.

Mereton, Philip. *The Heaven at the End of Science.* Distant Drum Press, 2009.

Meyer, Stephen. *Return of the God Hypothesis: Three Scientific Discoveries That Reveal the Mind Behind the Universe.* Harper-One, 2021.

Milonni, Peter. *The Quantum Vacuum: An Introduction to Quantum Electrodynamics.* Academic Press, 1994.

Milton, K. A. *The Casimir Effect: Physical Manifestations of Zero-Point Energy.* World Scientific, 2001.

Mitchell, Edgar. *The Way of the Explorer.* G. P. Putnam's Sons, 1996.

Moody, Raymond. *Life After Life.* HarperOne, 2015.

Newton, Michael. *Memories of the Afterlife: Life Between Lives.* Llewellyn Publications, 2009.

———. *Destiny of Souls: New Case Studies of Life Between Lives.* Llewellyn Publications, 2000.

Peña, Luis de la and Ana Maria Cetto. *The Quantum Dice: An Introduction to Stochastic Electrodynamics.* Kluwer Academic Publishers, 1996.

Penrose, Roger. *Cycles of Time.* Knopf Doubleday, 2011.

Rosenblum, Bruce and Fred Kuttner. *Quantum Enigma: Physics Encounters Conscious,* 2nd Ed. Oxford University Press, 2011.

Rubin, Peter. *Future Presence: How Virtual Reality Is Changing Human Connection.* HarperOne, 2018.

Russell, Peter. *From Science to God: A Physicist's Journey into the Mystery of Consciousness.* New World Library, 2004.

Smith, Huston. *The Religions of Man.* Harper & Row, 1986.

Talbot, Michael. *The Holographic Universe.* HarperCollins Publishers, 1991.

Targ, Russell. *The Reality of ESP: A Physicist's Proof of Psychic Abilities.* Theosophical Publishing House, 2012.

Taylor, Steve. *Spiritual Science: Why Science Needs Spirituality to Make Sense of the World.* Watkins Publishing, 2018.

Thomas, Andrew. *Hidden in Plain Sight: The Physics of Consciousness.* 2018.

Tierney, John. *Our Lives, Controlled from Some Guy's Couch.* New York Times, Aug. 14, 2007.

Tipler, Frank. *The Physics of Christianity.* Doubleday, 2007.

Tompkins, Ptolemy and Bernard Haisch. *Proof of God: The Shocking True Answer to the World's Most Important Question.* Howard Books/Simon & Schuster, 2017.

Virk, Rizwan. *The Simulated Multiverse: An MIT Computer Scientist Explores Parallel Universes, Quantum Computing, The Simulation Hypothesis and the Mandela Effect.* Bayview Books, 2021.

Walsch, Neal. *Tomorrow's God: Our Greatest Spiritual Challenge.* Atria Books, 2004.

———. *Conversations with God*: Books 1, 2, 3. G. P. Putnam's Sons, 1992–1997.

Watts, Alan. *The Book: On the Taboo Against Knowing Who You Are.* Vintage Books, 1966.

Whitworth, Brian. *The Physical World as a Virtual Reality.* CDMTCS-316, Research Report Series, December 2007.

Acknowledgments

Since our book is co-written, we will start off speaking jointly.

We want to thank each other for thirty-six years of successful marriage and a creative partnership in many areas of expertise: as co-authors, as co-brainstormers for Jovion Corporation (*www.jovion.com*), as co-managers of the *Journal of Scientific Exploration*, and as songwriters. Together, we have raised three children and managed a household. In short, we are a team, and each of us is grateful for the support of the other.

We thank our children—Kate, Taylor, and Elizabeth—our son-in-law, Jason Brenneman, and our two grandsons, James and Will Brenneman, for their love and support.

In addition, we want to thank those who shared their encouragement, comments, and feedback as we wrote this book—Bill Gladstone, our agent, Michael Pye, Stephan Schwartz, Larry Dossey, Eric Priest, Ervin Lazlo, Stephen Post, Professor Peter Sturrock, and Hugo Trux.

We are grateful for our long-time relationship with the Society for Scientific Exploration, because it inspired us to open our minds to new ideas and to think outside the box. The opportunity for Bernard to act as editor-in-chief and for Marsha to act as executive editor for ten years put us on a pathway of discovery. During this time, we interfaced with creative scientists and others who sparked many new ideas.

And we thank Garret Moddel, George Hathaway, Al Abel, Hal Puthoff, Dean Radin, Russell Targ, Jane Katra, Frances Kermeen, Gangaji, Frederick Winthrop, Eliza O'Malley, Michael Moran, and Verismo Opera.

Marsha also wants to thank Bernard for his unwavering support of her talents and encouragement of her singing and music teaching.

Bernard wants to thank Marsha for her caregiving skills and unwavering support of his scientific, metaphysical, and other creative endeavors.

Index

About the Authors

Astrophysicist Bernard Haisch, PhD, is the author of over 130 scientific publications, and was a scientific editor of the *Astrophysical Journal* for ten years. After earning his PhD from the University of Wisconsin in Madison, Haisch did postdoctoral research at the Joint Institute for Laboratory Astrophysics, the University of Colorado at Boulder, and the University of Utrecht, the Netherlands. His professional positions include staff scientist at the Lockheed Martin Solar and Astrophysics Laboratory; deputy director of the Center for Extreme Ultraviolet Astrophysics at the University of California, Berkeley; and visiting scientist at the Max-Planck-Institut für Extraterrestrische Physik in Garching, Germany. He also served as editor-in-chief of the *Journal of Scientific Exploration*.

Prior to his career in astrophysics, Haisch attended the Latin School of Indianapolis, the Saint Meinrad Seminary as a student for the Catholic priesthood and received a Bachelor of Science degree at Indiana University.

Marsha Sims, MM, has a multi-disciplined background and years of experience interfacing with business, supporting scientists and their projects, and immersing herself in the music world. She has a bachelor of music degree from the University of California, Santa Barbara, and a master of music degree from Notre Dame de Namur University. She has served as administrator and department secretary at Lockheed Palo Alto Research Laboratory; executive editor at the *Journal of Scientific Exploration*; administrator at California Institute for Physics and Astrophysics; and as administrator, graphic artist, and photographer at ManyOne

Networks and Digital Universe Foundation. She has taught voice, piano, and guitar at her own company (*www.marshasimsmusic.com*) since 2009. Marsha has performed over twenty lead roles in community theater and opera in the San Francisco Bay area. She was a church choral director and soloist for many years. She is also a certified Huna teacher.

Marsha and Bernard are a husband-and-wife team and work on many projects together, including songwriting (over one hundred songs), performing together in a few operettas, teaming up on the energy science company Jovion Corporation, and writing this book. Bernard and Marsha live in the San Francisco Bay area. They share three children—Katherine, Taylor, and Elizabeth—and two grandsons, James and William.